Urban Survival Guide

The Ultimate Step-By-Step guide for creating your Urban Survival Plan

David Morris

Printed in USA
SurviveInPlace.com
2274 S 1300 E #G15-192
Salt Lake City, UT 84106

Warning – Disclaimer

The information contained herein is the result of the authors' exhaustive study and research for his personal edification. Since individuals and situations vary, this information should be viewed solely for information purposes and applied with that in mind. You could die, get hurt, or end up in jail from applying the lessons in this book. Always get proper training and assistance from a qualified professional in your area before doing anything.

None of the parties involved in the production or distribution of this volume take, claim or accept any responsibility whatsoever for the use or misuse of the information contained herein regardless of motivation or intent.

Table of Contents

Introduction

This book was originally written as a 12 week course where lessons were written for delivery of the next lesson every seven days. Over 2000 students have gone through the course in this manner at the time of this printing and the positive feedback has been overwhelming. From career "door kickers" to people just waking up to preparedness, the course has changed many lives.

Each lesson of the course was broken up into an "easy to attack" format that could be completed at a pace comfortable for many different levels. Most of this formatting is retained in this book. Please consider each chapter as a lesson with exercises and assignments that can and should be completed at your own pace.

This is not to say that you cannot read through the book in its entirety and then go back to complete the lessons, but the original intention was a fully fledged Urban Survival course that builds upon itself. You will find more value in following through the book in this manner.

Research tells us that you will retain 9 TIMES as much of the information in this course if you complete the exercises. Put another way, you will only retain 10% of what you read for more than 7 days, but you will retain as much as 90% of what you read if you involve all five senses in the learning process.

When you add in the fact that those numbers will get cut in half when you're under stress, you need to ask yourself whether you want to remember 5% of the life-saving information in this book (reading only) or 45% (reading AND doing.)

Since you might be reading this book in preparation for a life threatening situation, I encourage you to take it seriously and DO the exercises. This course will be an invaluable tool for you, both for the peace of mind it gives you immediately, and the valuable logistics you'll have in place if you should ever need to use the lessons learned.

The lessons that you complete are designed to benefit you regardless of whether you ever find yourself in a survival situation. They are practical, pragmatic, and you will find yourself applying the lessons you're about to learn on a daily basis.

You'll note that I move through topics fairly quickly. I could get into detail, but doing so would stretch this course out for 100 or more chapters. I would lose most of my readers, and their survival plans would suffer as a result. The goal is to give you QUICK concrete, usable facts, procedures, and urban survival skills that you can use in the shortest amount of time possible. There are links at the end of each lesson to expanded resources. Due to the changing nature of our society, this is the best way to give you the most up-to-date information.

One of the goals of the course is to help as many people as possible prepare themselves so that they do not become victims of disaster or refugees subject to "efficient" treatment by government entities. The disasters could be natural or manmade, but the common element is the resulting breakdown in civil order that happens when people get hungry, thirsty, and desperate.

When we go over a particular area that you want me to cover in more depth, please post it in the members' forum and let me know (you can sign up at: www.UrbanSurvivalGuide.com/hardcopy). In many cases, I'll be able to direct you to a resource that I've used, or if there is enough demand, I can do an interview covering the topic, or I might do an entire course on the topic in the future.

We won't delve into the specifics of nuclear attacks, dirty bombs, EMPs, nerve agents, hurricanes, earthquakes, or any of the 50+ probable terrorist attacks that I know of through open source intelligence. Doing so would, again, make the course prohibitively long. I WILL give you general urban survival skills that you can adapt to use with ANY of these situations, as well as many more.

If you want to discuss a particular topic that I haven't covered, please go to the members' forum at www.UrbanSurvivalGuide.com/hardcopy

A large portion of my audience will be married men in the US. Many of you will not fit this demographic and may be single, a single parent, living alone, with friends, relatives, or any number of other arrangements in locations around the globe. I understand that, but in order to keep the course readable, I'm going to primarily refer to a husband/wife setup in the US. The course will work regardless of your situation.

I will also be including humor (free of charge). This course is very serious, but the topic is also very FUN. Movies, books, and TV series have been devoted to survival because it is an entertaining and engaging topic and I don't want you to be so bogged down in preparing for disaster that you miss out on enjoying life in the meantime.

I interviewed more than 30 subject matter experts for the SurviveInPlace course and this book. Almost none of them want their names to be released. There are a couple of reasons for their desire to remain anonymous. First, by remaining anonymous, they were willing to let their guard down and be more honest with me. Second, most of them keep a low profile and have neighbors who don't really know how "high speed" they are...and they want to keep it that way. I'm incredibly grateful for their wisdom, their willingness to help with this project, and the lives that their contributions will save and enrich.

Finally, you're probably wondering why I wrote this course and book. To make a long story short, there are three main reasons:

1. I have always been "prepared" and well trained, but in early 2008, I realized I had SEVERAL major gaps and began taking serious steps to fill them in. I realized that our particular family situation made bugging out of the city a very unlikely solution, and I needed to figure out how to survive short/long term breakdowns in civil services and/or civil order without leaving the city. So I started contacting friends and contacts in the Special Operations community, law enforcement, combat medics/paramedics, military contractors, intelligence operatives, chem./bio experts, Pentagon assets, and more.
2. As a result of my research, friends and family wanted to tap into my research and I had to figure out an efficient way to organize the information to give it to them. Many questions are answered in this book as a direct result of their feedback.
3. I truly believe that the future of a neighborhood, region, and even the entire US depends, to a large extent, on how many self-reliant, prepared people there are when adversity strikes. A case in point is the yearly response of Midwesterners to flooding compared to the response in New Orleans to Katrina.

There is a resource page at the end of the book (page 173) with web page resource pages for each lesson as well as a URL to sign up for the members' forum and other resources.

Make sure to keep up to date by visiting our forum and blog at SecretsOfUrbanSurvival.com. You should also sign up for weekly email updates while you're there.

You can also stay in touch through Facebook and Twitter:

Facebook.com/SurvivalDave

Twitter.com/SurvivalDave

Urban Survival Guide

Chapter 1. Overview

In this chapter, we're going to define urban areas, go over some quick reasons why it's good to get out of urban areas if possible, why most people will need urban survival skills, creating your SurviveInPlace™ Plan, operational security, and getting your family to buy into the preparedness mindset.

What is an urban area?

For the purposes of this book, urban areas are any area where the density is high enough that people use a shared water and/or sewer system…another way to look at it is that if you live closer than a quarter mile from your neighbors, you're in an urban area. That being said, the lessons in this book have been very helpful to people who live in rural areas just outside of urban areas…in other words, it WILL help you in your situation.

Why you should leave the city:

Your best chance of survival in a disaster situation will almost always be in a rural area. Cities have too many factors going against them to be ideal survival locations:

1. Cities are a target rich environment for terrorists.
2. Terrorists can blend in more easily in a city than in a small town where everyone knows each other.
3. Hazardous materials are shipped by rail and truck through cities.
4. Chemical plants and refineries are located in/near cities.
5. Gangs.
6. International airports (easy spread of disease).
7. Bacterial and viral infections spread easily in highly populated areas.
8. People are removed from their food and many don't know how to get food that doesn't come from a store.
9. A typical US city has a 9 meal (3 day) food supply, at which time all of the food is gone and it has to be resupplied from an outside source.
10. People haven't had to prepare as a way of life.

11. The density of prisons, criminals on probation, criminals on bail, former criminals, and criminals released on ankle monitors.
12. The entitlement mentality is more accepted in urban areas. (You owe me).
13. Overworked, underpaid, undertrained, understaffed police departments.
14. Hospitals are not staffed or equipped for disasters…thank God they aren't, or medical services would cost even more than they do now.
15. Many urban dwellers depend on others to take care of them…it's common for women to get comfort from carrying their cell phone as they cross a deserted parking lot at night, even though no criminal would stop an attack simply because they knew the police might be coming in a few minutes.

In a disaster situation, all of these factors will be multiplied. One in particular is police departments. I know several very respected officers around the country who in a disaster situation, would choose to stay home and protect their wife and kids instead of going to work and risking their life for you and me. And, to be honest, I can't blame them.

Why your best choice might STILL be to SurviveInPlace™.

Packing up and "getting out of Dodge" may not be a solution for everyone, and I'm guessing you relate to that. I get a little tired of survival experts saying that the only way to survive is to get out of the city. You know what…that just isn't possible for everyone. Some quick reasons why it might be better for you to SurviveInPlace™ are:

1. You've got relatives who are unable or unwilling to move due to age, sickness, physical shortcomings, finances, denial, or baseless "Hope". It could easily be a no-brainer for you to choose the high risk situation of staying with them rather than the fear of abandoning them to an unknown fate.
2. You may not be able to find a job that you're trained for in a rural area, or you might just be in a financial pickle where you can't afford to leave.
3. If you or a loved one are undergoing long term medical treatments and need to be close to a particular hospital for insurance purposes.

4. You might have a retreat to go to, but miss the window of opportunity to leave town and now the roads are choked with out-of-gas cars.
5. One of your immediate family could be traveling on business.
6. You or your spouse might be a public servant (police, fire, EMS, military, or utility) and you are duty bound to stay.
7. You don't have anywhere outside of the city to go.
8. The "disaster" is a deadly virus outbreak and you can just ride it out safely at home.
9. You experience a natural disaster like an earthquake, volcano, mudslide, or tsunami.

I could go on but you get the point. The reason that I took so much time on this topic is that you are going to talk with people at work and at gatherings who say that the ONLY way to survive a disaster is to leave the city. That's the ideal solution, but it's just not for everyone and you need to know that you're not going to be the only good person left in town when disaster strikes.

Folder/Notebook/Note cards for your SurviveInPlace™ Plan

This course is VERY much hands-on. The key to your success is going to be thinking about the given topics and organizing your thoughts enough so that you can write down a logical plan. You can write as much as you want, but you will not HAVE to write very much for any given exercise.

As a note, if your family is not intimately involved with the creation of your SurviveInPlace™ Plan, you will need to write with a little more detail so that they can follow the plan you are taking the time to create, but don't let writing out the "perfect" plan keep you from moving forward through the chapters and putting a "workable" plan in place.

What I suggest is that as you go through the course initially; allocate 95% of your time to thinking through scenarios and coming up with workable solutions. Spend the remaining 5% writing down the solutions in a way that YOU can understand them. As you have time, go through your SurviveInPlace™ Plan with your family and edit/clarify as necessary.

You're going to want to keep your SurviveInPlace™ Plan organized and you're going to have the following items to keep organized:

1. Printouts of the chapters/lessons (not applicable as you have the book).
2. Hand written notes/plans
3. Computer printouts of notes/plans
4. 3x5 Note cards
5. Article clippings
6. Catalog clippings

It's important that you get started immediately, so don't wait until you have the "right stuff" to organize everything. If you've got a folder and some printer paper, just use that to start with. We have our personal SurviveInPlace™ Plan set up in a 1 1/2" binder with tabbed sections. I use a 3-hole punch on printouts and written notes & have a few folders for 3x5 cards, loose notes, and clippings.

Operational Security

I want to QUICKLY touch on Operational Security. Operational security is the concept that you need to keep your battle plans secret so that your enemy won't find them out and gain an advantage over you. This is a VERY real issue with survival planning since you never know how your neighbors are going to act when they are out of food and their children are hungry.

The joke among my **close** friends is that their disaster plan is to get to my house as soon as possible. I'm actually counting on it, because they can carry their own weight in a survival situation and we can use the extra eyes, muscle, firepower, and brains.

But I don't want a bunch of people from work, church, my neighborhood, or even friends who haven't prepared at all to show up at my door. I'd have no choice but to turn them away creating ill-will or give away food/supplies that may mean the difference between life and death for me or my family.

The problem with giving someone food once is that you have only fed them once & they will come back to you again…likely soon and with other hungry friends. I intend on being helpful and charitable, but I will do it on MY terms, which include being anonymous.

Some things to avoid now so that you don't have these issues before or during a disaster. You want to:

- Avoid putting NRA type stickers on your car, as well as wearing obvious gun caps/shirts. (Almost ALL people with NRA stickers on their vehicles have guns in their house, just waiting to be taken when they're gone.) Gun shows & the gun range are good exceptions.
- Avoid painting a truck camouflage colors if you live in the city.
- Avoid public conversations about preparedness, like: "I've got all the guns, ammo, food, & bullion I need, but I really need a safe!"
- Avoid showing your survival supplies to anyone you don't trust completely.
- Avoid keeping your survival supplies all together where someone can see exactly how much you have.

In fact, if you happen to have a year of food & 100 gallons of water, a perfectly honest reply to someone asking you how much you've got would be that you have "enough food for a week or so and a few gallons of water. How about you?" You've answered the question without making yourself a target.

Just remember that parents will go to extremes to provide for their children.

If your neighbors' kids are hungry 10 days into a disaster that has no end in sight and they have seen your year's supply of food, you can guarantee that they're going to do whatever they need to get food for their kids.

If they think you're just barely making it, you're going to be much safer than if they think they can force you to give up your food.

You always have the option of putting food on their front porch anonymously.

The situation is going to be different if your neighbors are all on board with you and you have equivalent levels of preparedness, but there's no harm in underplaying your level of preparedness to all but your inner circle.

TO DO: Come up with a toned down answer for when people ask you what preparations you have made for disaster. Practice saying it until it is natural. Write it down in your SurviveInPlace™ Plan.

If you already have survival supplies in place, think about how visible they are and who comes in your house:

1. Friends
2. Friends of your children
3. Plumber/AC/Electrician
4. Babysitter
5. House sitter
6. Housecleaner

Assume they have no discretion, don't appreciate OPSEC, and will remember what they saw if they are ever in a survival situation. More to the point, your children may have friends who currently have drug habits or who might have a drug habit in the future. Assume they will remember anything that they see.

TO DO: Walk all through your house with the eyes of a stranger and see if your survival preparations are obvious. If they are, either camouflage your supplies immediately or make a written note in your SurviveInPlace™ Plan to make the necessary changes.

Get Your Family On Board

A. Don't force survival on them.
B. Use the local, national, and worldwide news.
C. Feed (don't smother) their growing interest.

One of the worst things that you can do if you want to get your family on board with survival planning is to force the topic on them. After all, do YOU like it when people ram an idea down your throat?

A much more effective approach is to use events in the news (any source will do) or even books, TV shows, and movies to start "what-if" discussions.

As an example, there's an alarm company ad on TV where a home invader kicks open the front door while the wife is home alone. He runs away when the alarm goes off, the wife calls the alarm company for help, and all ends well.

This brings up SEVERAL questions that you can ask:

1. Do you set the alarm when you're home alone?
2. If so, does the alarm go off immediately when the door opens, or is there a delay so you can punch in your code?
3. Does a loud noise (hitting two pieces of wood together) set off the alarm?
4. How does your alarm company respond to an alarm? (time, call, drive-by, call the police)

Personal Note: Our glass-break alarm went off when we were on vacation once and it took 20 minutes for our alarm company at the time to call us! They sent someone out to "look around" and that took another 15 minutes. They couldn't call the police just based on the alarm going off. In the city we were living in, a PERSON had to SEE and report an issue before the police would respond. Once the police got the call, it would have taken a few more minutes for them to respond. LESSON: Know your alarm company's response time and response protocol.

5. Ask your spouse if "calling the police" is a realistic strategy. Have your spouse play the role of the bad guy and see if you can get to a phone, dial, and connect to ANY number before your spouse gets to you. (Don't call 911...call someone on speed dial, like your home number or your spouse's cell phone.)
   a. When your spouse is able to EASILY get to you before you connect the call, say, "let's try it again...what should I do differently?"
   b. If you have kids, address how your plan would be different if they are/are not home.
   c. You can do this drill in EVERY room of the house, but don't drag it out longer than your spouse is willing to.
   d. Use this opportunity to identify improvised weapons in whatever room you're in.

e. One of the best home defense strategies if you're alone is to RUN out another door, screaming "RAPE!" (if you have close neighbors)

Make sure NOT to be critical of your spouse's suggestions. It is much better to say, "That sounds like it could work…what could we do to make it better?" or "I think that would work…what do you think about grabbing a couple of knives from the butcher block, screaming, and charging right at him?" (Don't encourage your spouse to practice stabbing you with real knives.)

Every time you finish one of these learning opportunities, make sure and ask a question that is more or less, "Don't you feel better knowing that you have a better handle on that situation if it ever happened to you? Also, try to boil it down to the essence. It could be Break-in-run-scream-get help. Or Break-in-closest weapon-attack-stop the threat-get better weapon-call police.

TO DO:

1. **Go through as many of these scenarios with yourself, your spouse, and your children as they are willing to do every day this week. (max 2-3 per day).**

2. **As you go through these scenarios and break them down, make a note of them on 3x5 cards and keep them with your SurviveInPlace™ Plan with the scenario on one side and the solution on the other. Review and update them as necessary until your family has internalized them. If you don't have 3x5 cards yet, just tear a piece of printer paper into 4 pieces and use that.**

3. **Get in the habit of looking for and identifying improvised weapons, wherever you are.**

Necessary for future chapters

Starting with next chapter, we'll be identifying danger zones in your city, potential assets, choke points, and travel routes. It will be very helpful to have a detailed map to make notes and draw on.

TO DO: Buy a detailed map of your city. The first map you should buy is a foldable map. In addition, you can buy a "book-style" map, but it is not necessary.

Now, before I mention next chapter's assignment, let me say this –

This is an "overview."

It is intentional. It is purposeful. I've arranged it this way so both beginners and experienced folks can get started immediately and get a general idea of what we'll be doing over the next several weeks. During the next few assignments, I'll give beginners baby-steps to complete each of the things we've talked about so far and I'll also provide some powerful insights for our advanced folks, especially in Chapter 6-10.

Now, on to our assignments…

<u>This Chapter's Assignment</u>

To Do:	**Date First Completed:**
Come up with a toned down answer for when people ask you what preparations you have made for disaster. Practice saying it until it is natural. Write it down in your SurviveInPlace™ Plan.	
Walk all through your house with the eyes of a stranger and see if your survival preparations are obvious. If they are, either camouflage your supplies immediately or make a written note in your SurviveInPlace™ Plan to make the necessary changes.	
Go through as many scenarios with yourself, your spouse, and your children as they are willing to do every day this week. (max 2-3 per	

day with your spouse and children, unless they ASK to do more).	
As you go through these scenarios and break them down, make a note of them on 3x5 cards and keep them with your SurviveInPlace™ Plan with the scenario on one side and the solution on the other. Review and update them as necessary until your family has internalized them. If you don't have 3x5 cards yet, just tear a piece of printer paper into 4 pieces and use that.	
Identify at least 3 items that you could use as a weapon in every room in your house.	
Buy a detailed map of your city. The first map you should buy is a foldable map. In addition, you can buy a "book-style" map, but it is not strictly necessary.	

Chapter 1 Resource page:
www.urbansurvivalplan.com/595/lesson1

Chapter 2, The Will To Survive

One of my good friends was the head of the SERE (Survival Evasion Resistance Escape) program at Offut AFB for several years. One of his stories is so important, that I'm including it here.

Back in the 80s, there was an incident where an F-16 pilot needed to make an emergency landing and landed at an abandoned airstrip in Alaska.

He landed perfectly. The plane was unharmed and he was fine. Unfortunately, he thought that his distress signal did not get out, and gave up, pulled out his pistol, and shot himself.

They estimate that he did this within 30 minutes of landing.

Help arrived within 2 hours of the initial distress call, which would have been well before his water/food/heat or any other supplies ran out.

This is a common story. In wilderness situations, people often die after a single night of "exposure" at 50-60 degrees, even when they have proper clothing. Soldiers who have watched too many movies showing instant "kills" from gunshot wounds have died in Iraq & Afghanistan after receiving otherwise non-lethal injuries.

People simply give up.

On the other side, one of the more amusing survival stories is of a gentleman who crashed his plane in a desert area and survived for almost a week in extreme heat/cold with almost no supplies, skills, food, or water.

The driving force for his survival?

He was in the middle of a divorce and refused to die and let his wife get everything.

The point of this is that the mind is a VERY powerful tool, and will either be your worst enemy or your most valuable tool in a survival situation. There are two easy steps you can take to make your mind work for you.

1. Choose to have a positive mental attitude.
2. Have something bigger than yourself to live for.

Entire books have been written on this topic, and if you want some suggestions of authors/books that I've found particularly helpful, please check out the resource page for this chapter. To be honest, this was an area that I had to address myself when I started going through the process of fixing my survival plan.

I was so focused on the bad political, economic, and global social events that were going on that I had stopped practicing the discipline of thinking positively, regardless of the situation.

In short, if you haven't already, you need to make a discipline of thinking positively. This doesn't mean that you walk around with rose colored glasses on or ignore reality, but it does mean that you control your mind. You still need to acknowledge when problems exist, but focus on finding solutions and what it will feel like to have successfully navigated the situation.

There's a famous saying, "Who by worrying can add a single hour to his life?" that is very true. Over the last few years, many people have been worried about a flu pandemic of one sort or another, terrorist attacks, or economic collapse. How does "worrying" about it help you? How does worrying hurt you?

Besides affecting your brain chemistry negatively, hurting your ability to sleep, making you depressing to be around, and increasing blood pressure, it wastes time.

A better approach is to only concern your mind with things that you have control over. As an example, you don't have any control on whether or not there is a global flu pandemic.

You do have control over how you are/are not going to respond if it becomes a reality and effects you. Identify the threat, figure out your plan and move on. By going through this course, you're going to do just that...create logical responses to potential threats so that you don't have to waste your life worrying about things that may or may not happen. You'll address them head-on once, write out your plan for dealing with them, and then go on living your life.

You're also going to need to have something bigger than yourself that keeps you moving forward.

Take a few seconds right now and imagine a Mad Max scenario. While I don't know if we'll ever see a time like that, it provides a good mental image for this exercise.

So, you're in Mad Max world and it's TOUGH. You've lost friends and loved ones. You don't have AC, a bed, or showers, let alone food or clean water. Your time is divided between avoiding danger and providing basic necessities.

Why would you keep going? Wouldn't it be easier to just roll over and quit?

NO!

But you have to train your mind to believe this, and the earlier you start telling your mind why it needs to keep going in hard times, the better it will respond when the time comes. So what are your reasons?

God? Family? Creating a safe place for your children to grow up in? Some cause?

For myself, it's that I want to do more for God and that I want to protect and spend more years with my wife and son.

I've got several other things that are important to me. Friends, relatives, the concept of liberty, etc, but God, wife, & son are the 3 things that I would choose over all others, including myself.

The point is that they have to touch you emotionally at a deep level and need to be things that are worth going through pain to protect/preserve.

TO DO:

Complete the following sentences for as many items as you honestly can:
"I would walk through fire for...."
"I would walk through fire to...."
Put the written list in your SurviveInPlace™ Plan.

If you need to, read the course description at www.SurviveInPlace.com. As you go through it, some sections will resonate with you and cause you to think of the people/causes that you would walk through fire for.

Please make sure and complete this exercise. It will not only help you in a survival situation, but by identifying what is most important to you now, it will help you complete this course and successfully put your survival plan in place. It might also cause you to make some major life changes like moving closer to family or finding a job where you work less than you do now or in a field that you're more passionate about.

Communicating with family during an emergency:
In a very strange way, people were much more mentally prepared for communication breakdowns before cell phones became popular.

If you look around any public area, you'll probably see half of the people around you with a phone up to their ear and most of the rest will have a phone in their lap, on their waist or a visible Bluetooth headpiece. 15 years ago, if any of your relatives wanted to get a hold of you, they'd just have to wait until you were at home or your office. Now, they EXPECT to be able to reach you immediately and not being able to reach a friend/loved one quickly can be cause for panic.

15 years ago, I actually knew the numbers for my family and friends. I didn't have speed dial, and I could call everyone who was important to me from any phone. In fact, I still remember the phone numbers of many of my grade school friends. I had new numbers written down and sometimes carried a day planner with the rest of my numbers.

In those days, I could fall in a pool without losing all of my important numbers. In an emergency situation, I could get soaked by a fire sprinkler system and still find a phone and call my relatives to coordinate a time/place to meet up.

Two years ago, I realized this had changed...I carried numbers in my wallet, but I didn't know my own home phone number, or any of my family members' cell phone numbers. I didn't have to because they were all in my phone, but what that meant is that if my phone died or didn't work in an emergency, I couldn't get in touch with any of them.

My wife and I have made and carry a business card that we printed out on our printer using Avery printer card stock (Avery 8878). They have our basic information (name, blood type, medical information) on them along with key contact numbers. This allows me to have a written backup for the key numbers I would need in an emergency and it's easy to occasionally pull out the card and refresh my memory on the key numbers. Since the realization that I didn't actually KNOW my important numbers anymore, I've started taking out this card occasionally and memorizing the information on it.

TO DO:

Write/type your important information and contact numbers on a business card. Start carrying it with you and memorize the information on it.
Print it out on water resistant stock if possible (Avery 8878) but don't wait to take this step. Use an old business card of yours or someone else's and simply write on the back of it until you have business card paper to use.

When NOT to call after an emergency.
Recently, when I was in San Francisco for a "geek" conference, I learned a valuable lesson about how cell phones work during emergencies.

I was downtown at the Moscone Center with 20,000 other geeks and it seemed like at least 1/3 of the room had iPhones. My phone was working fine initially, and then I couldn't get email or browse anymore. Then when lunch time came around, I couldn't get any of my friends on the phone, even though text messaging worked.

The phones worked sporadically throughout the rest of the day. IF I was able to connect with anyone, we'd lose the call within seconds, even with a strong signal.

Curious, I called AT&T later that night and found out that the sheer concentration of data enabled phones being used at our conference caused the closest cell tower to overheat. This put more of a load on the surrounding towers and they crashed like dominos. By the time AT&T had the problem identified and figured out, 7 towers had burnt out circuitry and txt messages were the only traffic that were getting through during the day for the next 2 days.

I researched this more when I got home and found out that this also happened at a recent Apple conference and it happened in New York on September 11th, 2001. (The urban legend is that Blackberrys are the only phones that work in a disaster. While it is correct that Blackberrys worked on 9/11/01, the complete truth is that the network that Blackberrys communicated on was the only one working. Blackberrys now use the same network as other phones and texting is the best way to communicate when calls won't go through.)

The lesson here is that **if you find yourself in an emergency situation and start losing calls, stop trying to make calls and switch over to JUST texting.** You'll save your battery, and most likely will be able to communicate quicker than calling.

Some carriers have announced plans to turn off data and/or SMS messaging in disaster situations, so it's wise to have a backup plan to use for communication instead of a cell phone.

Common Long Distance Contacts

There have been some recent disasters in the US (hurricanes, tornados, floods) where the phone lines were so loaded that local calls could not be completed. Even so, long distance calls still went through. The takeaway here is to have a primary, secondary, and tertiary person that everyone in your family can call/text in the event of an emergency who lives in another area code than you do.

Everyone in the family can call into these numbers, leave and retrieve messages, and communicate their location and situation.

Prioritized List of "Stuff" to Buy

On the resource page for lesson 2, I've included a link to a spreadsheet template that you can use to start keeping track of purchases you want to make, their cost, & their priority. There is an example at the end of this chapter.

There are two major reasons for taking this step:

1. It will help you take care of needs before wants.
2. It will help you save money

Since you don't know when a local, regional, or national emergency will happen, buying survival items in the wrong order could leave you with some major gaps in your plan.

As an example, let's say that you're at Costco and you see a generator that you really want, and it's on sale for $100 off. That's great, but fortunately, you remember that you only have 72 hours of emergency food. It's pretty obvious in this case that the money would be better spent on their 275 meal Bucket-o-food for $85.

If you don't have a prioritized list that you've thought through of survival items that you need/want, you're going to end up with a mish-mash of survival supplies with some elementary gaps.

My personal prioritized list is based on categories and goes like this:

1. Shelter (tarp/foil blanket/tent/sleeping bag/etc.)
2. Water (boxed/bottled water/iodine tablets/Chlorine/boiling/filter/purifier)
3. Fire (matches/lighters/lenses/flint-mag/"high-speed" tools/skills to make fire)
4. Food (cupboards/stored food/self-reliant sources)
5. Medical (Including prescriptions necessary for survival)

After these big items are taken care of, THEN take care of the following:

Security (dogs/weapons/skills/training/alarms)
Comms (wireless/wired/local/extended range)
Extra clothing
Comfort
Barter (cash/gold/silver/.22/cigarettes/liquor/Imodium/iodine tabs)

What I encourage you to do is to create a list of all of the "stuff" that you want to buy. Next, go through it and designate whether or not it is something that you absolutely need or something that you just want.

For right now, just put in the items that come to mind immediately and print it out. You'll find that you are continually adding to it, re-prioritizing, and changing what you want as you do more research and do more. It's fine to make most changes in writing and occasionally update the spreadsheet version.

Also, put in a time component. In other words, figure out what you would need to survive for 72 hours in each of the 5 categories above. This can be taken care of very inexpensively.

Next, go back through the 5 categories, but this time figuring out what you'd need to survive for 7 days. Then extend it out to 6 or 12 months.

What you'll end up with for food is multiple items like this:

Survival food for 72 hours
Survival food for 6 months
6 month supply of food we currently eat

In this instance, you could first get your fast & light survival food, then, get a 6 month supply of rice/beans/oatmeal, and finally start buying double of the foods you buy now. We'll cover food storage in the subsequent chapter, but this strategy of combining what you currently eat with food storage has several advantages over simply going out and buying a pallet of "survival food."

You can download your template by going to www.surviveinplace.com/spreadsheet and export it as the appropriate format for your spreadsheet software (.XLS works for Microsoft Excel) or you can save it to your online Google docs account, if you have one.

Once you have your prioritized list filled out, send it to me at SIPBook@surviveinplace.com. I'll pick 1 list a week to personally review and make recommendations on. I'll post this on the site, so everyone can benefit from it.

Insulin, Heart Medication, & Pain Meds

Since we're addressing real life scenarios, let's take a look at medical conditions. If you have a medical condition that requires you to take medication multiple times a day to survive, then you're going to want to consider getting a prescription for 3-6 months from your doctor and buying the medication with your own money.

This will be prohibitively expensive with some medications, but it is an issue that you need to address. If you HAVE to have a particular medication every day or you'll die or stop functioning within a week, then it doesn't make much sense to purchase survival supplies at a faster pace than you're stocking up on your medication.

Another alternative is to go to Chicago and visit www.NaturalHealthCenter.mercola.com. In most cases, they can show you how to change your diet so that you can safely get off as many medications as possible. Their approach is unique and worth checking out.

The second major advantage of having a prioritized survival list is that when you see "sales" & "specials", you'll know whether or not they're a good deal and you'll be able to make quick decisions. Just like using a grocery list when you go to the store, you'll also find that you buy more of what you need and less of what you don't need.

This week's BIG exercise

This week's big exercise involves turning off your water/power for an evening. It will be eye opening for some and old-hat for others. I grew up in the middle of the country in tornado/blizzard country and have lived through a few tornados and several multi-day blizzards. In fact, we normally had at least one 3-7 day storm each winter that isolated us and knocked out our power. If you live in a similar area, I still encourage you to do this exercise…if for no other reason than it's fun!

There are 3 parts to the exercise:

1. Learning how to shut off your utilities.
2. Preparing your freezer for power outages.
3. The actual exercise.

If you don't have access to your utilities (you're renting, live in a high-rise, etc) then turn off everything that you can and find out what you could turn off if there was a situation where you needed to. As an example, if you live in an apartment and there was a major gas leak, you might want to turn off the master breaker as you're evacuating the building.

I need to tell you that you are responsible for your actions. Electricity, gas, and water can be very dangerous. They can damage your valuables, destroy your house, or kill you if you do anything wrong while working with them. Please consult your local expert before doing anything with water, electricity, or gas or before doing anything suggested in this training.

After you have consulted with your local experts, make sure you know how to turn off your electricity, water, and gas if you have it. The process will be different depending on whether or not you get your power from a utility, are off the grid, have municipal water or well water, and whether you use gas, oil, or propane.

Gas

If you have natural gas, I suggest that you don't turn it off, as it can create a dangerous situation with some systems and you may have to relight the pilot lights on your appliances.

Do make sure that you know how to turn it off and that you have a tool to do so. You might even want to turn your gas shutoff valve 1/8th of a turn to make sure it is not stuck. If you don't know what you're doing with gas, please consult with someone who does.

Water

If you don't have a water shutoff where your plumbing enters your house, you may need a water shutoff T-wrench or "Water Meter Key" to turn it off. They range from a couple of feet long to 10 feet long, depending on how deeply it freezes where you live. You can get them at your local hardware store and they cost less than $20. Ask someone at your hardware store to demonstrate how to use it. It's VERY easy.

Exercise Part II: Preparing Your Freezer

A full freezer will keep meat frozen for approximately 2 days if it's full, but only one day if it's only partially full. It is also more efficient because it's less affected by frequent openings/closings.

What we do to take advantage of this is take plastic water bottles from the store and put them in our freezer. (If you don't know whether or not a particular bottle will break/leak when frozen, put it in a plastic bag) This is particularly nice in the summer. When we go for a drive or a hike, we just reach in the freezer and grab these frozen bottles & replace them with warm ones. We use them as ice blocks, and have the added bonus of nice cold water to drink as the ice melts.

A couple other alternatives to water bottles are to use Nalgene bottles ¾ full (so they don't break as the ice expands), ice cube trays, or a household plastic pitcher filled part way.

How much water should I put in my freezer?

It depends on how full your freezer is. The purpose of the water is to keep the freezer cold longer in the event of a power outage, so if you already have a full freezer, then you don't need to put any water in it. Also, if you don't have anything in your freezer that will spoil if it warms up then you don't need to put any water in it, although it's not a bad place to store some water.

Part 3: Jump in!

Pick one of the next few nights and turn off your electricity and water for the evening. If you've got kids at home, make it a fun event. Call it "Family Game Night", or "Camping At Home" or any other creative name that you come up with. Remember, most survival preparations are pretty darn fun!

Here's why this is important. If you're going to put your faith in a plan to survive for 3, 6, or 12 months, you'd better make sure that it works for a single evening.

By going through this simple exercise you'll find out several important things. Do you know how to cook without a stove? Does your emergency stove work? If you don't cook, **are your emergency rations edible**? Do your lamps / lanterns really light a room so you can read or play games? Do they stink too bad to use anywhere other than camping?

Make sure to write down what works and what doesn't work. In particular, pay attention to toilet/hygiene issues, food, cooking, light, & games/activities.

Here are some helpful hints for this exercise:

1. Pick an evening when your entire family will be home.
2. You may have to do some creative negotiating with your family to buy into this exercise and a Friday/Saturday night may be the only night that will work.

3. Start at 5PM or whenever everyone is home.
4. End as late as possible. If you have battery powered alarm clocks, you can even wait until morning.
5. Don't use your gas appliances.
6. Turn off your phones/laptops/iPods/game systems etc. (FYI, cordless phones won't work without power.)

You may very well find the silence of a "powered down" house to be very relaxing, especially if you have ADD. It also means more distraction free time with your family. With no electronic noise, you can actually talk to each other, enjoy games, and read books.

Small assignment:

As you're traveling this week, make a written note of the following:

1. Concentrations of graffiti or gang activity.
2. Streets/roads/intersections that get congested easily.
3. Railroad tracks.
4. HC (Hazardous Cargo) routes on highways/interstates.
5. Locations of chemical plants, refineries, & fuel storage.
6. Other threats & dangers in your area.

Next week, we'll go into this in more detail and I'll show you how to find more of this information online and we'll add it to your map so you can have a strategic view of your city.

Review of This Week's Assignments

To Do:	Date First Completed:
Complete the following sentences for as many items as you honestly can: "I would walk through fire for…." "I would walk through fire to…." Put the written list in your SurviveInPlace™ Plan.	
Write/type your important information and contact numbers on a business card. Start carrying it with you and review it often so that you will eventually memorize it. Print it out on water resistant stock if possible (Avery 8878) but don't wait to take this step. Use an old business card of yours or someone else's and simply write on the back of it until you have business card paper to use.	
Create your prioritized list of Survival items to buy or trade for. Put the written list in your SurviveInPlace™ Plan.	
Make sure you can turn off your utilities	
Add water bottles to your freezer.	
Spend an evening/night with the utilities turned off. Write down what worked well and what issues you need to figure out before spending 72 hours without utilities.	
Start writing down potential dangers and congestion areas for your area.	

Example template for prioritized list of Survival supplies:

Item	Date	(N)eed/(W)ant	Category	Priority	Cost	Store	URL	Date Purchased

It is obviously too small to actually write in these boxes…I've included it as a template for you to follow to create your own prioritized list. You can download your own at: www.surviveinplace.com/spreadsheet

Categories:

- Shelter
- Water
- Fire
- Food
- Medical
- Security
- Comms
- Clothing
- Comfort
- Barter
- (add your own)

Chapter 2 Resource page:
www.urbansurvivalplan.com/592/lesson2

Chapter 3. Current Potential Threats

In addition to talking regularly with local law enforcement, you're going to want to be observant of whether or not criminals are currently active in your particular area.

As an example, if you start seeing a big increase in graffiti in your neighborhood, it could indicate gang activity; or it could simply be "taggers". If you want to learn more about gang graffiti and how to tell the difference between simple vandals and gang members, I suggest that you go over to Robert Walker's site, "Gangs Or Us" quickly and watch a 2 minute video on gang graffiti deciphering. For now, just watch the 2 minute video and bookmark the page so you can go back when you're done with this chapter.
www.gangsorus.com/graffiti.html

You can also view all of the crimes that your local law enforcement is reporting by going to one of these two sites:
www.crimereports.com

If your city does not share data with crimereports.com, try:
www.spotcrime.com

On crimereports.com, I suggest clicking on the "Crime Types" button and selecting violent crimes. Then, pick the 30 day option.

This will quickly show you all of the violent crimes committed in your area in the last 30 days, as reported by your local law enforcement.

In the upper right hand corner of the map, it will tell you how many crimes are being shown. Crimereports.com will show a maximum of 500 crimes on a given map, so make sure you zoom in so that this number is under 500.

The addresses that have a teal blue box are addresses where multiple incidents have happened in the last 30 days. If there are areas with multiple teal blue boxes, consider marking them on your map.

Accidents / Terrorist Targets / Natural Disasters

Take a look at your map or ww.nationalatlas.gov/natlas/Natlasstart.asp if you don't have your map handy and locate the railroad tracks in your area. If you are within a mile of the tracks, you need to have a plan in place for rail accidents.

One such incident happened on January 6th, 2005 at 2:40 AM when two trains collided in Graniteville, South Carolina causing 90 tons of chlorine gas to be released into the air. Since it was the middle of the night, there were 5,400 residents sleeping within one mile of the accident and of those 5,400 residents, 9 people died of chlorine inhalation and 250 people had to be treated for chlorine exposure.

Before you get too worried, I want you to put this in perspective. There are only 5-10 rail crashes a year in the US and the Graniteville crash is the WORST one since 2005! Most deaths in rail accidents happen because of the actual crash and not because of chemicals released into the air.

That being said, in addition to accidents, Arabic terrorist websites have been promoting the use of homemade thermite to cut rail lines or weld debris to rail lines in urban areas to derail trains suspected of or known to be carrying hazardous cargo. These sites are also promoting attacks on industrial chemical storage facilities. In particular, they're targeting anhydrous ammonia (fertilizer), hydrogen fluoride (pharmacology, plastics, refineries), methyl isocyanate (pesticides, adhesives, rubbers), and chlorine.

What's that mean? It means that if you live/work near a rail line, you should know that a crash/toxic leak is a possibility and decide what your response would be if a large scale chemical release happens. It isn't likely to happen, but if it does, you may be the only person in your immediate area who has thought through what to do and who can act immediately and guide others around you.

Fortunately, your preparations for this scenario will also carry over to the more mundane (and likely) threat of an industrial fire upwind of your location.

As an example, if you hear a massive train crash ½ mile upwind from you during a "normal" time and immediately see a cloud rising up into the air, you might want to take action by getting out of the area.

Let's assume that there was no explosion and you can figure that if the wind is blowing straight towards you with a light breeze (10 mph), you probably have 3 minutes before any of the smoke/chemicals reach you and the best course of action is to leave in your car and tell your co-workers/neighbors to do the same.

Let's say that you are directly north of the explosion. If possible, you want to escape directly to the East or West so that you will be completely out of the path of the cloud. If you remember your geometry, you want to put distance between you and the incident while traveling perpendicular to the direction of the wind.

Keep in mind that your response will most likely be different if a large scale chemical release happens during "normal" times than if it happens during a local/regional emergency when normal travel isn't an option.

If such an event happens during a time of general emergency, or if you aren't one of the first to evacuate, you may have to Survive In Place™ . We'll be covering this in a couple of weeks, including how to create a small safe area in your home or office within minutes that will allow you to drastically increase your chances of surviving a chemical incident, should one happen near you.

You should also consider developing a similar plan if you live within a mile of a chemical manufacturing company, refinery, fuel storage facility, or manufacturing facility that has large stockpiles of chemicals.

Start making note of these facilities as you're driving around and don't be afraid to ask firemen if there are any facilities near your house that you should be aware of.

You can also get the locations of many of these facilities right now by using Google Maps.

Simply go to maps.google.com and enter the following search:

Refinery Houston (or whatever your city name is)

And press the "search maps" button.

You'll get back results for actual refineries and the offices of refineries. A general tip is that if a result is in the middle of the city, it's an office and if it's in a large open area, it's a refinery. Google has the option to do a satellite view or street level viewing. Both will let you see whether or not there is an actual refinery or not, but since you're only concerned with facilities that are in your immediate area, another option is to simply drive by and see what's there.

Another search to do is:

Chemical dealers Detroit

When the results come up, look above the results on the left hand side and you'll see the following:

Categories: Chemical Dealers

Click on "Chemical Dealers"

Most of these companies don't have a large enough quantity of chemicals on hand to be concerned about. If you have any within a mile of your house or where you work, drive by some day and see if they have large outdoor tanks. If they do, mark them on your map. If you don't know whether or not they're a danger, just put a "?" on the map. If you know what the chemical is, go ahead and write it on your map.

This can be especially important if you live in an earthquake prone area. If you have an earthquake and need to get home from your place of work quickly, you would do well to avoid facilities that may be leaking irritating/harmful chemicals into the air.

One last thing to be aware of if you live within ¼ mile of an interstate is hazardous cargo routes. Like railroad accidents, hazardous cargo accidents don't happen very often but they involve much smaller quantities of chemicals.

Next time you're driving into your city, look for square signs that either say HC or HC with a line through them. They are in place to direct truck drivers on the legal routes to get through a city with hazardous cargo.

This section may make it sound like there are threats all around you…and there are, but to put it in perspective, please go to the following site:

www.hazmat.globalincidentmap.com/ and click on the "HERE" button in the lower right hand corner.

You can see a map of all of the hazardous materials incidents in the US for the last 30 days. You'll see that the number of true chemical (non-fuel) incidents is incredibly small.

There are more gas spills on highways and meth houses then there are large scale chemical accidents.

I want to bring this to your attention so that you don't waste undo time/money preparing for a large scale chemical release that probably won't happen or affect you.

For most people, it's enough to know that the threat exists, work through a plan so that you know what to do if you see/hear an accident or are asked to evacuate.

Civil Breakdown

Hurricane Katrina showed us firsthand what can happen in the US in an urban area after a major disaster when order breaks down. We're going to cover survival lessons from Katrina in depth, but for today, I just want to go over one thing: Avoid large groups of people in need. In particular:

1. Stadiums
2. Homeless Shelters
3. Schools
4. Auditoriums
5. Any area designated as a Red Cross Shelter
6. Hospitals
7. Churches

Yes, even churches. After Katrina, many churches were forced by the city to provide services to whoever the city brought to them. This was fine when the churches had food, toiletries, and the plumbing worked.

It wasn't fine when the plumbing stopped working and the city wouldn't let the churches send people home. It got ugly, many crimes were committed in churches, and many churches were horribly vandalized.

Unless you're actively working at one in a relief capacity, you'll be better off avoiding the area around all of these facilities. Why? Well, it's where panicked people trying to get in will be. It's where people looking for handouts/victims will be hunting for their next mark. It's also where people who got kicked out of the facility for fighting/drugs/etc. will be congregating, and drug dealers will be peddling to refugees from the shelters.

Your area may be different. In many parts of the Midwest & the Rockies, church members wouldn't stand for a situation like this and churches will be a place of refuge during/after an emergency. If an emergency happens in your area and you decide to go to a church, listen to your gut. If it doesn't feel safe as you're approaching or while you're there, leave immediately.

Mark these facilities on your map, but only the ones in areas near where you are likely to be when a disaster happens.

Choke Points

You're also going to want to mark down choke points that could keep you from traveling between your work and your house, from your house to someone else's, or from your house out of your city.

What I mean by a choke point in this case is an area that is susceptible to traffic jams and likely to slow you down or stop you.

You'll know many of these immediately areas from your daily travels. What roads/intersections get backed up during the morning commute? The afternoon commute? On Fridays before a 3 day weekend? What stop lights take 3 lights to get through?

There's another category of choke points that will also be an issue in disaster situations, and those would be bridges and canyons where an accident shuts down all traffic, places where the number of lanes of traffic goes down, and construction areas.

On a personal note, sometimes when my son can't sleep, we go driving around the city until he does. On one such night, I decided to drive one of our routes out of the city to see if there was anything new to be aware of. I was more than a little surprised to see that a big stretch of it was under construction and was now 1 lane instead of 3!

Construction choke points are short term, so I wouldn't mark them on your map, unless you want to buy a new map every 6-12 months. Just be aware of them.

You also want to be aware of military installations. In the event of a terrorist attack or civil unrest, it's very likely that roads that go parallel to military installations will be closed to create an additional safety buffer.

Skill/Weakness Self-Assessment

Today, we're going to fill out a Skills/Weakness assessment so that you will know what survival skills you have, which ones you want to work on, and which ones you want to look for in other people.

Take 10 minutes and fill out the skills assessment for this chapter. You can find it on the resource page for this chapter and there is an example at the end of the chapter as well. Be as honest as possible with your self-assessment and make sure to fill out the last column. If you know multiple people who work for a particular skill, write them all down. These are people that you want to have a relationship with and have pre-planned with before a disaster happens.

Don't worry, we will be covering strategies for approaching these "highly skilled" people in a future chapter without compromising OPSEC.

If you're really serious about this assessment, have somebody close to you assess YOUR skills. If there's a big discrepancy between your rating and theirs, it could mean that you are over/under confident, or that you simply need to show them an example of what you can do.

Basic Supplies Inventory

This will take you between 15-30 minutes, depending on whether or not you're able to do it all at one time or not.

This is NOT a comprehensive survival list, it is an initial inventory of basic items that you should have for Surviving In Place in your house.

Take a notepad and take a QUICK inventory of the following categories of items:

Fire
Food
Water (and low sugar, no caffeine drinks)
Fuel
Batteries
Trauma Supplies
Medicines (expiration dates)
Vitamins
Prescriptions

This inventory does not have to be exact, and can look something like this: (Note, the following is not what I suggest you have on hand and is only to illustrate what you should write down)

Fire:
Fireplace & 1/2 cord of wood
Camp stove
Propane grill
Propane single burner
12+ boxes of matches
12+ lighters
Flint/magnesium fire starter
20+ candles

Food: (non-perishable food that you won't be eating in the next 7 days)
40 cans of soup/vegetables
5 pounds of potato flakes
275 serving Costco meal bucket
2 3600 calorie bars
60 cliff bars

Water:
60 gallon water heater
5 x 32 oz Gatorade
24 x 1 liter bottles of water

Fuel:
3 x 5 gallon Unleaded (Dated)
2 x 20 lb propane
1 gal white gas (camp fuel)
400g isobutane (camp stove)

Batteries:
20 AA
10 AAA
2 9V
0 C
0 D
3 Lithium flashlight batteries
1 extra watch battery

Trauma:
Band aids from mid 1990s
Costco first aid kit

Medications:
Advil: 300+
Tylenol: 225
Aloe: 16 oz
Imodium: 16 oz

Vitamins:
60 days

Prescriptions:
36 contacts & 1 pr glasses

You can also include ammo and other survival supplies that you have on hand, but again, this exercise is meant to cover the most basic items. Over the next few weeks, we'll go over what you eventually want to have on hand, what you want to have in your 72 hour kit(s)/go bag, and the most cost effective way to take care of it.

Review of This Chapter's Assignments

To Do:	Date First Completed:
Update your map with threats and choke points near and between your house and place(s) of work/school/relatives.	
If your home/work is vulnerable to a large scale chemical release, decide on a response.	
Assess your Urban Survival Skills/Weaknesses	
Take a basic inventory of your survival supplies	

This is a Skill/Weakness assessment to help you see where your strengths are and what weaknesses you need to fix, either by learning new skills or by finding local people who have strengths that are complementary to yours. Be honest in your assessment. It will help you to pick the right people to team up with in the event of a survival situation. As an additional step, make a copy and have someone close to you fill out the assessment rating their impression of your skills.

Skill/Attribute	How You Rate Yourself	Who's The Best You Know
Essential		
Building a Fire (using matches or lighers)		
Water Purification		
Medical (emergency)		
Important for Urban Survival		
"Hard" Skills		
Plumbing		
Electrical		
Automobile Repair		
Mechanical		
Electronics		
Radios		
Ropes/Knots		
Medical (health/wellness)		
Levers/Gears/Pullys /mechanical systems		
Empty Hands Fighting		
Improvised Weapons Fighting		

Knife/Stick Fighting		
Tactical Handgun		
Tactical Long gun		
Tactical Shotgun		
Survival Skills (experience with survival hardship)		
Military Training		
Camping/Hiking (general skills)		
Fitness		
"Soft" Skills		
Negotiating		
Barter		
Encouragement		
Problem Solving		
Administration (organizing things and thoughts)		
Leadership		
Spiritual (Peace in the storm)		
Creative thinking (improvising)		
Coolness under stress		
Relationship Counseling		
Able to push through failure		
Positive Attitude		

Chapter 3 Resource page:

www.urbansurvivalplan.com/590/lesson3

Chapter 4. Purpose of the 72 hour kit

In this chapter, we're going to cover a core survival item: 72 hour kits. Depending on where you live, you may have heard them called bugout bags, GO (get out) bags, blizzard kits, emergency response kits, or any of a number of other names. If you've got your basic inventory of survival supplies from last week handy, you'll be able to complete much of the process of creating or upgrading your 72 hour kit with the essentials in a matter of an hour or less.

72 hour kits are primarily designed to help you survive a broad range of emergencies for 72 hours away from your home or other sources of shelter and supplies. The purist assumption is that you won't have anything other than the 72 hour kit and items that you acquire. This could be because you are stuck on the side of a road, because you had to abandon your car away from civilization, or because you had to leave your house and were only able to grab one thing. It could also be because a storm or chemical/biological event stranded you at your office overnight or longer.

There are also dozens of secondary uses for 72 hour kits. Over the years, I've used my 72 hour kits for many practical reasons besides "emergencies" such as:

- Being at a trailhead and someone "needs" toilet paper.
- Having a runny nose and nothing to wipe it with.
- When I can't stop to eat, but have low blood sugar.
- I cut my hand changing a tire and needed to clean/bandage it.
- I decided to spend the night at a friend's house & wanted to brush my teeth.
- I couldn't get home because of a blizzard.
- Car broke down & I wanted food/water while waiting for a tow.
- We ran out of diapers in our diaper bag.
- Being at a picnic with charcoal, lighter fluid, and no matches.

Again, I want to stress that emergency preparedness doesn't have to be all doom & gloom and I encourage you to thoroughly address situations that are more likely to happen (like a short-term regional natural disaster or a simple accident) before you get concerned about stocking your 72 hour kit with items for long term civil breakdown or an economic collapse.

There are literally hundreds of items that you would want in your 72 hour kit if you were preparing to get dropped into a Mad Max scenario, but we're going to focus on developing solid basics today. We'll also talk about optional items that you'll want to consider in your 72 hour kit.

I want to start right off by saying that there is NO perfect 72 hour kit. Why? Because the perfect 72 hour kit would have to meet AT LEAST the following criteria:

1. Inexpensive
2. Lightweight
3. Small
4. Handle natural/manmade emergencies in 4 seasons
5. Allow for young children, the elderly, guests, sick, or injured people
6. Contain 3 days worth of shelter, water, fire, food, security, tools, comfort items, medical items, & clothing. (A gallon of water weighs 8.3 pounds)

As you can see, "perfect" is not a reality. Fortunately, "great" is.

72 hour kits are very personal items and you are going to have to make the final decision as to what you put in yours, but I'm going to give you some general guidelines to help you.

Logistics:

You're going to have to make some decisions about what kind of 72 hour kit(s) you are going to have. How much weight can you carry? How much are you willing to spend? How many people are you going to prepare for? How much room do you have in your car? How many kits are you going to have? What disaster is most likely to affect you?

In fact, let's answer those questions right now:

How much can you carry?	
How many kits are you going to have? (1 per car + house + office?)	
How much are you willing to spend?	
How many people are you preparing for?	
How much room do you have?	
What disaster is most likely to affect you?	

Depending on where you work, you may want a 72 hour kit in your office in addition to your car. This is especially the case if you work in a high-rise building above the 4th floor or if you commute/ride to work.

If you travel for work, you may want a modified 72 hour kit. I fly a few times a month and even have a TINY, light, bare bones 72 hour kit that I carry in my checked baggage. (I also carry it when I go on long trail runs or fast & light hiking)

Even if you work from home, you're probably going to want to have multiple 72 hour kits. As an example, we have purchased several one-size-fits-all 72 hour kits for testing purposes, but three of our kits are primary ones. We have one in each vehicle and one in the house. The ones in the cars are designed to support two adults and our toddler for 3 days and the one in the house is designed to support all 3 of us at a higher standard of comfort for 3+ days and has room to add last minute items. The car ones are small (2400 cubic inches…like a school book bag) and the house one is a full size 7000 cubic inch internal frame camping backpack.

We actually use our big 72 hour kit every time we go camping. In addition to knowing that everything in it works and that I know how to use it, this serves as a great way to rotate items and always know that everything in it is ready for immediate use.

Although there are several configurations available for 72 hour kits, including a 5 gallon bucket, duffel bag, box, (we have ALL of them), a sweatshirt sewn shut, etc., I suggest using a backpack or having one available for the simple reason that they're easier to carry.

A good rule is to cover all of the essential items initially and then expand out from there. In other words, unless you find an incredible sale, make sure you have basic shelter, water, fire, and food covered before you get a GPS or Chem/Bio/Nuke items.

You also want to make sure that you know how to use everything in your kit, THAT IT WORKS, and that your body can survive on it. One of the 72 hour kits that we bought had a multi-tool in it already. As I was going through the kit, everything was awesome...solid stuff. I almost passed up the multi-tool thinking that it would be as good as everything else, but fortunately I didn't.

To begin with, I couldn't get the darn thing open without pulling my finger nail away from my skin. That was enough to toss it, but I also quickly found out that the metal was soft and completely worthless for any task other than holding paper.

So make sure that you try everything that you might trust your life to.

I'm going to break down 72 hour kits into essentials, basics, recommended items & optional items so that you can go at whatever pace your budget allows. This is the order in which you want to stock your kits:

1. Essentials in your home 72 hour kit.
2. Essentials in your car 72 hour kits.
3. Basics in your home kit.
4. Basics in your car kits.
5. Recommended items in your home kit.
6. Recommended items in your car kits.

Essential items:

The Essentials are going to cover one of the 4 tenants of survival, which are Shelter, Water, Fire, & Food. You should have most of these items on hand to put in your 72 hour kit already, even if they're VERY basic.

Shelter is anything that will protect you from the elements, including heat, cold, wind, rain, and sun. The items that will work as shelter will depend on where you are, the season, your budget, the number/age of people you're preparing for and how much weight you can carry. Remember that this is a 72 hour solution, so you can use a space blanket, poncho, bivy sack, tarp, a tent, heavy plastic, or contractor grade garbage bags and duct tape.

We consider our car our primary shelter when we're away from home, but carry space blankets, ponchos, and a tube tent in our car kits in case we need to abandon it. In our home kit, we also carry a space blanket & poncho, but have a heavier tarp, & nicer tent.

Water can be difficult to find or make drinkable in an emergency situation, especially in urban areas. You've got a few options for your 72 hour kit:

Method	Weight	Cost	Pros	Cons
Carry it	8 lbs per gallon	A few dollars	Time/convenient	Weight/space
Chlorine drops	A few ounces	pennies	Cheap/light	Finding water / taste/time
Boil	Depends on your system	Pennies	simple /warms the body	Finding water /uses fuel/ time
Tablets	A few ounces	A few Dollars	Light/cheap/ no fuel	Finding Water / Taste/time
Filter	Around a pound	Up to $200	Fast / taste	Finding water/can clog/must keep parts separate
Exotics (miox or UV)	A few ounces	Around $100-$200	Fast / taste / light	Takes batteries / can break

These are all going to have varying levels of effectiveness depending on how much sediment is in the water, and what you're trying to get out of the water. Your purification is going to be different depending on whether your water has bacteria, viruses, salts, herbicides, pesticides, fuel, oil, acids, bases, or other urban contaminants.

As a general rule, the more cloudy and smelly the water is, the more careful you need to be before drinking it. If you have the time, it's a great idea to let water sit for an hour or more to let particles settle out and pre-filter your water with a t-shirt, handkerchief, or dedicated pre-filter to remove sediment.

I cover some really cool methods of getting salts, oil, and other urban contaminants out of water in a mini-course at UrbanDisasterWaterPurification.com. For now, we're going to focus on the main water contaminants that you face on a daily basis, which are viruses and bacteria. This would be applicable for non-contaminated stream water and contaminated tap water after a disaster.

Again, we take a different approach for our car kit and our home kit. Our car kits have boxed water (which hasn't broken for us in freezing single digit weather), Katadyn purification tablets, and the ability to boil.

Our home kit also has boxed water, but has nicer filters, including a Sawyer inline filter (rated for 1MM gallons and a GREAT filter), a Katadyn inline carbon filter, purification tabs, a Katadyn base camp filter, and gear to boil water easily. The only reason we carry so many is because, with the exception of the actual water, they're all small and light, and have specific purposes.

If I could only carry one (besides boiling), it would be the Sawyer inline filter. Using two bladders (1 clean & 1 dirty), you can set up a gravity system that's fast, simple, and effective, or you can use it in-line with your current hydration pack. In addition, they come with a faucet adapter so that you can filter water from hydrants, hotel faucets, hoses, and any other pressurized source.

What water filtration/purification system(s) are you going to use?

Fire is the next most important item for survival. We carry the same items both in our car kits and our home kit, "waterproof" matches, 2 lighters, and a flint/magnesium stick. ALWAYS have a backup for matches, even "waterproof" or "survival" matches. I can't over-emphasize the importance of carrying simple lighters. They light hundreds of fires, dry off quickly and if you get a "torch" style one, it's even windproof. They're small and cheap enough that it's silly not to carry a spare or two.

Fire has many benefits in a survival situation, including heat, light, comfort, cooking, purification, signaling, and mental health. Even a fire that is ineffective as far as heating can "warm the spirits" and make a cold night more bearable.

We also carry a 100 hour candle for light instead of wax candles. Why? Wax candles melt in a hot car in the summer.

Although they're not really fire, I'm going to include flashlights in this section. I am a very big fan of the newer BRIGHT LED lights that run on AA and AAA batteries. They're light, last a LONG time, and some even have multiple brightness settings. Surefires are my favorite, but they're expensive and use non-standard batteries. You can usually buy packs of two non-tactical LED lights at warehouse stores for $10-$20, including batteries.

We keep batteries in our flashlights that are in the front seat area, but we keep the batteries separate in our 72 hour kits so that the light won't get turned on and use up the batteries. It also protects the light from a corroding battery.

If you're concerned about not having immediate access to light in your 72 hour kit, you can put a small light stick or a keychain light on the outside or in a pocket where you can find it immediately.

Finally, we keep a shakable flashlight in our kit. When you get a shakable flashlight, make sure to use it somewhat roughly for a few days to make sure that it won't fall apart. We've got a few of these lights, all from China, but some hold up very well and others literally fall apart. The one piece of advice I can give you on selecting one is that, in general, the lights that feel more solid have performed better than the ones that felt "cheap."

What fire/light tools are you going to put in your 72 hour kit?

Food is the next item to have in your 72 hour kit. You not only want to have enough calories, but you also want to make sure that your system can handle what you're giving it and that you're getting minerals (especially electrolytes), and vitamins.

We carry several 1200 calorie bars, Cliff bars, and candy in our car. We've got the same in our home pack, and we also have a couple of freeze dried meals and 2 cans of food that I change out from time to time. (and a can opener)

Keep in mind that freeze-dried meals are light, but require water, fuel cooking, time, and are expensive. We like several kinds of freeze-dried/dehydrated meals, including Mountain House, Backpackers Pantry, and Natural High.

Another benefit of these meals is that you can "cook" in the package without getting a pot dirty. Simply boil your water, pour it into the package, and let it sit until the food is rehydrated and ready to eat.

We've also found the Costco 275 meal Bucket to be a good source of freeze-dried meals. It's $85 for 275 meals. The meals come in separate packages and each package has 5 servings. My wife and I can fill up on ½ of a package at a cost of 77 cents regularly and will use an entire package ($1.44) or add canned meat if we've been hiking/exercising.

Canned foods are going to have water in them and many are pre-cooked. The extra water counts towards your daily water intake, so don't look it as "extra" weight. In addition, you can cook canned food IN the can and not have any cleanup.

Our home 72 hour kit always has at least one of our camp stoves in it. We have an esbit solid fuel stove, a Primus Omnifuel stove, a Primus EtaPower stove, and a Jetboil system. They're all great, but I'd suggest the Jetboil personal cooking system for its simplicity, versatility, weight, and size.

We also carry meal replacement shakes in our car. Try to get ones that have REAL sugar in them as opposed to artificial sweeteners like sucralose, splenda, saccrin, or nutrasweet.

What are you going to carry for food?

One other essential item is to have 1-2 weeks of any medications that you must have to survive. Why so much? Because in the event of an emergency, you can make your supplies stretch and can likely find additional food/water, but specific medication will be very hard to find.

So, these are the essentials for your bare-bones 72 hour kit(s): Shelter, Water, Fire, Food. They're the absolute basics for survival, and you're going to want to get these items in your house kit and car kit(s) as soon as possible. Once you've accomplished this, we'll move on to the basics.

Basic items:

The basic items in your kit are going to include items for first aid, medical, hygiene, security, and tools.

First Aid: I suggest getting ANY medium sized soft sided first aid kit with bandages as a base. You can get a great basic kit from REI or online for $20-$30 and customize it as necessary. Make sure to check with your local EMT supply store to see what basic kits they sell. I've found them to be 20-30% cheaper than the chain stores. If you don't know where your local EMT store is, ask a fireman or EMT.

Make sure that you know how to use everything in your kit. It's worth it to get a few extra of everything so that you can try it out and know what everything is. You might find that the "bargain" band-aids that you bought don't stick at all, or that you need more practice putting butterfly bandages on someone while wearing gloves.

I'm a wilderness EMT and have added a few other items to all of my kits over the years:

Item	Number to Buy	Date put in kits
Electrolyte tablets/powder		
Benadryl		
Decongestant		
Imodium Anti Diarrheal		
Superglue (I prefer Crazy Glue, but you can buy Superglue in bulk for CHEAP)		
Maxi Pads for women AND for pressure dressings		
Butterfly Bandages		
Nitrile gloves (rotate yearly if you're in a hot climate)		
Dentemp for making temporary dental repairs		
Moleskin		
Burn pads		

Tools: You're going to want a couple of good solid knives and a multi-tool as a minimum. I recommend having both a locking foldable knife (Kershaw, MOD, and CRKT are my favorite folders) and a fixed blade knife like the USMC KaBar knives. Make sure and get a "full tang" blade. Full tang means that the knife is made of one solid piece of metal from tip to tail. Just to make sure I'm clear on this, I absolutely do not recommend getting a knife with a screw off tip for holding survival items.

On the multi tools, try to get one that has the tools that you use most often and make sure to try it out so that you know you can open/close it easily and safely.

If you have any car-specific tools for battery cables, belt clamps, allen/star bits, etc., you want to include them in your kit. Also, make sure that you know how to use your jack and that you have the tools necessary to change your tires. I mention the jack because many new cars have PATHETIC jacks that you should either completely replace or you may need to carry a metal/wood base in your car to use underneath it on soft surfaces.

If you're going to do any off-road driving or driving in deep snow, you should also carry a shovel with you.

Another broad application "tool" you can carry with you is duct tape. You can use it to make some car repairs, fix shoes, cover wounds, help in making a shelter, sealing your car, and dozens of other uses.

Finally, make sure you have a pair of heavy leather gloves.

We have smaller hatchets in our 72 hour kits, but they really aren't practical to use when you need to cut big wood for a fire.

If you anticipate using your home kit for camping, it may be worth it to put a good camp saw in your home kit. I carry a SaberCut saw. It is basically a chainsaw blade with handles at both ends. It's not EASY to cut through a tree with it, but it is the best compact solution I've found.

Security: I don't suggest keeping a firearm in your 72 hour kit, because of the cost and because it won't be secured in your vehicle. There is also a risk that if a State of Emergency is declared that your weapons will be taken. Some items that you can keep in your kit that you can easily use as defensive weapons that are easy to hide and/or unlikely to be taken are pepper spray and a long choke chain.

Comfort/Medical: A couple of items that will be worth their weight in gold if you need them are toilet paper and Maxi Pads. Some other comfort/medical items to consider:

Item	Number to Buy	Date put in kits
Toothbrushes		
Hand Sanitizer		
Baby Wipes >> If you can, buy in bulk from warehouse stores		
Spare Glasses >> $10 glasses www.zennioptical.com		
Sting kit		
Epi-Kit for severe allergic reactions. Requires a prescription.		
Asthma Inhaler		
Diapers		
Socks		
Underwear		
Prescription Medication		
Anything you're addicted to (Nicotine/caffeine)		

Once you have your kits stocked with these items, it's time to move on to the recommended items.

Recommended Items:

Item	Number to Buy	Date put in kits
Cord (550/551 paracord or parachute cord)		
N.95 mask (sold out in April 2009 for the swine flu scare. The REAL reason to carry them is for dust/debris after an explosion, earthquake, or building collapse.)		
Hatchet		
Extra Ammo for your primary weapon		
AM/FM/emergency radio receiver		
2 way radios		
Pen/pencil/paper		
Signal whistle		
Aluminum foil (for cooking, signaling, improvised cup, etc.)		
Handheld GPS		
Spare batteries		
Roll of Quarters		
Boonie cap		
Local map		
Compass		
Garbage bags		
Urban Survival Playing Cards from UrbanSurvivalPlayingCards.com		
Urban Survival Bandana from UrbanSurvivalBandana.com		
Paperback book/Bible		
Deet		
Full change of clothes		
Peanut M&Ms		
Phone card		
Zip lock bags		

With the exception of spare clothes, all of the items up to this point will fit in a small backpack.

The list of optional items that you can include in your kit is endless. As a quick example, if you are in a cold climate and regularly wear dress shoes or heels, you're going to want to have a spare pair of shoes/boots in your car that you can change into if you need to.

Operational Security

Depending on what kind of bag you put your 72 hour kit in, it may look very appealing through a window or when you're loading your trunk at the grocery store.

In short, tactical looks valuable. If I see a high-speed Maxpedition bag in your car as I walk by, I'm going to think, "gun, ammo, knives, & cool kit!" Unfortunately, cool camping packs face the same problem

One option is to use a red/pink/orange bag with the intent of it looking like a teenage girl's book bag or an EMT/CERT bag. This has the downside of making the pack more visible if you are in an evasion situation, but it also makes you look less tactical.

Better yet, you can put your entire 72 hour kit (pack) into a garbage bag or a plastic container, slap a strip of duct tape on it and label it, "diapers" or "Stuff To Donate" with a magic marker. That way you can have the pack you want and still not make yourself a target.

Chapter 4 Resource page:

www.urbansurvivalplan.com/588/lesson4

Chapter 5. Flu and Pandemics

This chapter is a biggie, and has generated quite a bit of interest with recent coverage of the swine flu. There is more writing and fewer exercises, and there are a few items that you're going to want to buy as soon as possible.

We're going to cover a lot of ground and I'm going to do my best to give you thorough answers while keeping the chapter readable and the pace fast. At the end, I'll direct you to some additional in-depth resources.

I've consulted doctors, naturopaths, a bio-weapons expert who studied under the head of the Soviet Bio-Weapons program, and a geneticist who specializes in viruses for this chapter.

Like you, I'm particularly concerned about whether or not the threat is real, how to avoid the threat, and how to deal with it if it becomes a reality.

To begin with, let's talk about what the flu is and briefly decipher what all of the letters and numbers associated with the swine flu actually mean.

Swine flu is a virus that affects the respiratory system and is caused by the type-A influenza virus. There are 5 types of flus and type-A is by far the most common and the only one we will address in this chapter, although the information is compatible with other types of the flu and many other viruses.

One of our seasonal flus is the H1N1 version of the virus and it is basically an adapted version of our normal flu that has genetic materials from the pig version of the flu.

Virus adaptation is a normal process and all flus adapt to one extent or another as a means of self-preservation so that they will be able to infect people who have developed natural immunity to a previous strain.

The H and the N are simply the first letters of two proteins that make up the flu virus. They are hemagglutinin and neuraminidase. Basically, it's easier to say and write A(H1N1) than to write influenza type-A(hemagglutinin type-1, neuraminidase type-1). "Swine Flu" is even simpler, but they all mean the same thing this flu season.

The numbers are different versions of the proteins, so the A(H1N1) virus is simply the type-A flu with the type 1 hemagglutinin protein and the type 1 neuraminidase protein.

There are 19 types of flus that have been identified and 12 types have caused human deaths. The A(H1N1) is the same version of the flu that caused the 1918 flu pandemic, but "same" is somewhat of an overstatement. While both versions of the virus were type H1N1, the 1918 version was much more aggressive than the current one has been so far.

That being said, this version of the H1N1 virus is potentially deadly because it carries a swine version of the H1 protein that is different than the human H1 protein that many people already have developed immunity for.

How does the flu form?

The origins of flus in general and the swine flu in particular are debatable, but the most accepted theory is that the virus lives normally in birds without causing them any problems. When birds are in close proximity to pigs, occasionally the virus will infect a pig.

As the pig's immune system fights off the virus, the virus changes genetically to survive and then the virus is able to spread from pig to pig.

This was the case with the H1N1 virus. It was fairly stable in pigs for several decades, passing freely from pig to pig, but with only a few isolated human infections.

Pigs and humans are both susceptible to type H1 & H2 viruses, so when a bird passes one of those types of the flu to a pig and the pigs are in close proximity to humans, the virus has a chance to make additional adaptations to survive in humans. The H1N1 virus did this in 1998 by combining the existing proteins from bird flu, swine flu, AND adding proteins from human flu.

This process can occur anywhere where humans are in close contact with pigs and pig waste. Two primary situations are near pork production facilities and in cultures where pigs are seen as part of the family and sleep in the same area as people do.

This is most common in Southeast Asia, and the chain of events that have caused many of our widespread flus is fairly straightforward.

In the springtime, farmers (some of who have contracted the flu from their livestock) in Southeast Asia travel and get together in large groups for livestock fairs, Chinese New Year celebrations, Easter celebrations in the Philippines, and other springtime holidays.

These large gatherings allow the flu to spread from human to human and get spread through the region and internationally. Typically, flu seasons are worse during rainy years because people spend more time indoors (away from UV light) and in closer proximity to other people.

Why was the 2009 flu originally called the Mexican Flu?

There are a couple of theories on this:

First, If you've noticed in your grocery stores, a lot of produce now comes from Mexico. This is not just a US phenomenon. China also buys much of their produce from Mexico and one theory is that that the swine flu was transferred from China to Mexico by Mexican farmers attending a produce conference in China early in the spring of 2009.

They may or may not have been symptomatic before spreading the virus.

Second, it's very possible that the 2009 strain made the jump from pigs to humans in Mexico.

The first identified patients (patient 0) with this strain of the flu were believed to be from La Gloria, a town of 3000 people and 15,000 pigs east of Mexico City. Almost 30% of the town suffered from flu-like symptoms and three young children died shortly before Easter, 2009.

Everything changed when people traveled to visit relatives for Easter. Within a week, people started coming down with A(H1N1) across Mexico, and soon, thanks to international travel, across the globe.

We may never know where this strain truly made the jump from pig to man. No country wants to take credit for it, and the cities and facilities in question have made it difficult for researchers to determine the origin for the exact same reason. In reality, it's fun trivia, but if you get the flu, it really won't matter whether it started in China or Mexico, so I urge you not to get too hung up on it.

In any case, the theories have parallels:

1. People living close to pigs/pig waste
2. The virus makes the jump from pig to people
3. The virus makes the jump from people to people
4. Infected people travel and spread the virus

A brief history of flu pandemics (& pandemic scares)

Just so you know what the words mean, pandemics are simply epidemics that cover a large area and affect a large percentage of the population. Epidemics are infections that spread rapidly among a population.

Said another way, pandemics are infections that spread rapidly among a population and cover a large geographical area. Note that it doesn't necessarily mean that the infection is deadly.

The 1918 flu pandemic is the most commonly cited flu, and there are some pertinent facts that are important today since they're both A(H1N1) strains.

1. It had up to a 2 week incubation period, but killed in as fast as 2-3 days of getting symptoms, which was too fast for the medicines of the time to help.
2. Although called the Spanish Flu, it's believed to have made the jump from pig to human either in China or on a US pig farm in Kansas and spread to Spain by US troops. (similar to the possible misnaming of the "Mexican" flu)
3. It's estimated that 500 million of the world's 1.8 billion people contracted the flu and it killed between 25 and 50 million, although some estimates are as high as 100 million.
4. Very few people died of the flu, but rather most died from secondary infections including Strep-Pneumonia.
5. The virus showed up initially in the spring, went dormant in the summer, exploded for 8 weeks in the fall, went dormant again, and re-emerged in the spring of 1919.
6. Normally, people with bad strains stay home and recover while people with mild strains continue their normal activities spreading the mild strain. In 1918, soldiers at war didn't have the option to stay home and spread the more deadly strain.

Modern application of this is people crowding doctors' offices and Emergency Rooms with flu-like symptoms. Some people will have nothing, some will have mild strains (and weakened immune systems) and others will have the more severe strain and spread it to the others who will be released to go back to their daily lives.

The 1957 flu pandemic was a type H2N2 from pigs and was much milder than the 1918 virus, killing only 2 million worldwide.

Interestingly enough, the H2N2 virus hasn't had a widespread reach since the 1968 flu season.

Testing equipment, medical reporting, and international communication had improved significantly since 1918 and there were a few key lessons learned:

1. Schools were the most effective incubators of the virus in the US and UK, infecting as many as 90% of the students.
2. The virus didn't spread in the US until school opened in the fall.
3. The virus died out by early December and in late January/early February 1958, it re-emerged in the elderly population.
4. Infection rates were highest among people who attended large indoor gatherings (school, events, etc.)
5. The staph-pneumonia mortality rate was as high as 28% and regularly killed people within 48 hours of being admitted to the hospital because the antibiotics of the time didn't start working fast enough.

The 1968 flu pandemic was a type H3N2 from pigs and was even milder than the 1957 pandemic, killing 1 million worldwide.

The main take away from the 1968 flu was that it's thought that one of the reasons it was milder than the 1957 flu is that it didn't reach its peak until Christmas 1968 when children were home on break, thus limiting the spread in schools.

The 1976 flu (non-pandemic) is an example of what many fear for future flu seasons. A massive vaccination campaign was launched in fear of a pandemic. The result was $1.3 Billion in lawsuits by people who were paralyzed by the vaccine, 25 deaths, hundreds of people who were struck with Guillain-Barré Syndrome and no pandemic or even an epidemic happened anywhere in the world.

Modern day factors

It's important to realize that the regular flu kills 1 million people every year and doesn't make the news. In addition, malaria kills 3000 people EVERY DAY and is not considered newsworthy either. Even so, your friends and co-workers probably aren't worried about malaria as much as swine flu, so let's take a look at how it spreads, some bad (popular) ways to fight the flu and some effective ways to prevent and fight the flu.

N.95 masks

I have N.95 masks, both in my house and my vehicles, but they're NOT for the flu. Rather, I have them to protect me from airborne contaminants in the event of a local natural disaster or terrorist attack.

Why don't I like N.95 masks for flu protection? To begin with, the flu virus is relatively weak and is killed quickly by UV (sun) light. People who wear masks normally wear them when they're outdoors, where the virus will be killed quickly by the sun. The time to wear the masks is indoors, but even then it doesn't make sense.

When a person with the flu sneezes, spittle stays airborne for up to 17 minutes indoors, carries the flu virus, and can be absorbed through the eyes. More importantly, 17 minutes gives the spittle a long time to travel.

Some also ends up on flat surfaces, doorknobs, pens, etc. and you pick up the virus when you touch any of those surfaces. Soon enough, you will use the same hands to rub your nose, eyes, or mouth.

One of the biggest reasons not to trust the N.95 masks is that the flu virus is smaller than the particles that N.95 masks are designed to filter.

As icing on the cake, the National Institute of Health doesn't even recommend that people wear them to protect themselves from the flu.

In short, most effective use of N.95 masks to stop the spread of the flu virus is to have people with the flu wear them, just like surgeons do when they operate to protect the people they're operating on.

Tamiflu & Relenza

I have to preface this by saying that the topic of Tamiflu & Relenza is very emotionally charged. I interviewed experts who have dozens of doses of Tamiflu on hand and I interviewed experts who would never touch it. I have attempted to cut through the emotion on the subject and just give you facts that you can take action on.

Tamiflu is an oral antiviral prescription drug and Relenza is an inhaled antiviral designed to kill the flu virus. You can buy them for $50-$200 per dose and they have the big benefit of shortening the average duration of the flu by a day and a half. Considering the fact that secondary infections are the biggest killer associated with the flu, this is POTENTIALLY a big benefit.

Unfortunately, they also have many downsides. To begin with, neither Tamiflu nor Relenza (another anti-viral flu drug) actually attack or kill the flu virus. They ONLY play defense. My geneticist friend described it this way:

Picture a guy in a boat tied to a dock. The dock is one of your cells and the guy in the boat is the flu virus. Once he destroys the dock, he wants to cut the rope and go find another dock to destroy. Tamiflus only job is to keep the guy from cutting the rope. It won't attack the guy (the virus), and once you're done taking Tamiflu, the guy is free to go back to cutting the rope and destroying more docks.

In order for Tamiflu to be effective, you also have to have a healthy immune system that can come in and actually kill the virus. We'll talk about strengthening your immune system in a little bit.

Tamiflu also has side effects that are similar to the flu, including nausea, vomiting, diarrhea, headaches, dizziness, fatigue, seizures, and coughing.

In fact the FDA has issued warnings and Japan banned the use of Tamiflu on children because users between the ages of 10-17 were reporting the additional side effects of delirium, panic attacks, hallucinations, and convulsions.

We're already seeing strains of the flu that are resistant to Tamiflu and as use increases, so will the number of Tamiflu resistant strains.

I mentioned earlier that shortening the duration of the flu had the possible benefit of decreasing secondary infections. Unfortunately, users of Tamiflu report a higher rate of secondary infections than non users.

My family and I will not be taking Tamiflu or Relenza. If you haven't made an educated decision on the subject, I strongly encourage you to research the subject so that you will be confident in your decision and can defend it to people who are on the opposite side of the issue as you, regardless of which side that is.

I'm not anti-drug, anti-big pharma, or anti-doctor. I'm quite thankful for all of them, and in fact, many drugs are simply vitamins, oils, or other substances taken from nature and repackaged.

That being said, I tend to like treatments with as few side-effects as possible, treatments with as little "Big Brother" record keeping as possible, treatments that are easy to find, treatments that don't require me to sit in a room full of sick people, and treatments that don't require me to pay someone to give me permission to take the treatment.

Proven ways to prevent and fight the flu

As I said earlier, a healthy immune system is a key to fighting off the flu and getting over it quickly if you do get it…even if you do take an anti-viral medicine.

Fortunately, there are several cheap, simple ways to strengthen your immune system to keep from getting the flu and fight it if you or a loved one comes down with it.

Amazingly, one of the most effective ways to keep from getting the flu is with vitamin D through sun exposure. For a fair skinned person, 15 minutes of sun exposure without sunscreen will give you 10,000-20,000 IU compared to the RDA of 400 IU! (As an example, you get approximately 100 IU from a glass of milk)

How's this work? Well, vitamin D is converted into the hormone, calcitriol in the kidneys and increases the body's production of antimicrobial peptides. These peptides destroy the cell walls of bacteria, fungi, and viruses, including the influenza virus.

It's important to note that not all vitamin Ds are created equal. Most supplements are vitamin D2 but the form of vitamin D created by the body when exposed to sun is vitamin D3. Both are converted to their more usable form in the body, but D3 converts 500% faster and binds with proteins more effectively than D2.

The effectiveness of Vitamin D3 in fighting viruses has been studied since the 60s and the link between the winter solstice and the flu was first put forth by R. Edgar Hope-Simpson in 1981. Sunlight (Vitamin D3) is believed to be the main reason why the flu is worst in the fall/winter in stormy years among people who spent the majority of their time indoors.

The connection was made stronger when researchers studied the Inuit. Their diet consists primarily of fish (high in Vitamin D3), and they stay healthy through the winter months, despite less sun exposure.

One of the possible reasons that the 1918 H1N1 flu was so deadly is that it caused cytokine storms, which can be oversimplified as the body's immune system over-reacting and attacking the body and liquefying the organs. Fortunately, Vitamin D3 is a very effective drug for immune response in general and cytokine storm prevention in particular.

TO DO:

Make a point to start spending 10-20 minutes outside every day without sunscreen. While getting 10-20 minutes of sun exposure every day is the EASIEST thing that you can do to improve your immune system, it's also the easiest thing to forget to do or postpone. One strategy that will help you is to put an "S" on your calendar for the next 30 days. Every day that you get 10-20 minutes of sun exposure, cross through the "S".

IMPORTANT: I want you to realize that this entire course costs less than ONE dose of Tamiflu and you just learned a FREE way to dramatically lower your chances of catching the flu.

In an earlier chapter, I suggested contacting Joe Mercola's clinic in Chicago if you are on any long-term prescription medications. If you haven't yet, please do. You can go to his website at www.NaturalHealthCenter.mercola.com/

Joe is a licensed physician and surgeon who's focus is primarily on non-prescription solutions to medical issues. We don't always see eye to eye on political and non-medical topics, but Joe has helped me tremendously over the last 4 years with health and wellness.

Joe has not had the flu in over 20 years and has suggested 8 simple, inexpensive ways to dramatically improve your immune system.

1. Vitamin D through daily sun exposure
2. Avoid sugar and processed foods
3. Get enough rest (so you don't need caffeine to function) Getting less than 6 hours of sleep each night increases your chance of contracting illnesses by as much as 300%
4. Control stress (Prayer, meditation, and EFT)
5. Exercise
6. Daily omega 3 from fish, or high quality fish oil or krill oil
7. Wash your hands thoroughly throughout the day
8. Eat garlic often if you're able to

Additional strategies and dietary aids for you to consider for boosting your immune system and helping keep you healthy:

1. **Probiotics** will replenish the good bacteria supply in your gut and help you get more nutrients out of the food you eat. In addition, these good bacteria will help your body fight off bad bacteria. I'll include a link to the best source I've found on the resource page for this lesson.
2. **Echinacea** boosts the immune system by increasing white blood cell counts.
3. **Astragalus** is an effective herb to take to strengthen your immune system before getting ill, but should not be taken once you're sick.
4. **Stop Smoking.** I won't nag you on this. If you're addicted to nicotine, you're probably already trying to quit.
5. **Limit Caffeine Intake** to one glass of coffee per day. Once you get through the withdrawal, you will have more energy (not less), a lower stress level, and you'll be able to sleep better.
6. **Use a Paper Towel** when leaving bathrooms to open doors after you wash and dry your hands. Many places now have trash cans by the door for this purpose. If not, you can always carry the towel for a few seconds until you can throw it away.
7. **Carry a pen with you** to sign credit/debit card receipts rather than using the common pen at store registers.

I've got the flu. Now what?

Sometimes life throws you curveballs. You have a big project at work, a teething or colicky baby, family or financial issues, an injury, or something else that is outside of your control that affects your sleep and immune system. That's life.

If this happens during flu season and you do get the flu, don't panic! There are tried and tested ways to fight off infections that are cheap and effective.

Keep in mind that this is your life and/or the life of a loved one that you're dealing with. If this year's flu strain is killing people within 2-3 days of becoming symptomatic, I don't know how long I'd wait before trying to go to a doctor, and I'm not going to tell you how long to wait.

In fact, many "public advocacy" groups are working various media channels to try to get the message out that supplements and herbal remedies will not help you and that your only hope is to get to a doctor and buy Tamiflu.

The fact is, if 1/3 of your city is coming down with a killer flu, you may not be able to get an appointment with a doctor or be accepted to an ER, even if you want to. If you do get in, there may not be any prescriptions available.

Simply put, you can buy ALL of the items listed below for under $100. If you get the flu, want to go to the doctor, and are able to, then you haven't lost anything...you still have the items. They've got long shelf lives, and you can always use them in the future.

But, if doctors' offices and ERs aren't taking new patients or there is a breakdown in civil order, you'd better have another option ready to go.

I keep ALL of these items on hand, and my wife, myself, and our toddler use them when we get under the weather. Don't let their simplicity fool you. They are very effective.

I've got to tell you that I'm not a doctor and I'm not giving you medical advice. Consult your medical professional before doing anything in this book.

- In addition to continuing to get **daily sun exposure**, one of the non-prescription strategies is taking 10 milligrams of **zinc** daily from your multi-vitamin, lozenges, or sprays. If you feel nauseous after taking zinc, it's a good sign that you have enough in your system already.

- A simple and effective treatment for respiratory infections (like the flu) in general and pneumonia in particular is the use of high quality **oregano, thyme and rosewood oils**. You can put 5-10 drops of each in a warm bath, 5-10 drops of each in a steaming bowl of water and breathe in the steam, or 5-10 drops in a humidifier. It is important that you get the highest quality oils you can find. Inexpensive oils are usually only made to smell good and don't have the components that work for medical applications.

- A personal treatment that cured a case of drug-resistant pneumonia (contracted in the hospital) that nearly killed me, as well as other minor illnesses through the years is a combination of **Grapefruit Extract, Tea Tree Oil, and Colloidal Gold**. I'll include a link to the best source I've found on the resource page for this lesson.

- One very effective supplement for upper respiratory infections is **elderberry or sambucol**. It is a natural anti-viral and you can get it in either liquid or tablet form. You can even get Elder-Zinc lozenges that have both elderberry and Zinc.

- Again, **Echinacea** boosts the immune system by increasing white blood cell counts. Simply increase your dosage when you start getting ill.

- **Stop taking antacids!** The incidents of pneumonia are 4X higher among people who take Nexium, Prevacid, Pepcid, and Zantac than the general public. Stress was a considerable part of that equation, but current users still have an 89% higher chance of contracting pneumonia than former users.

Some of these treatments may be a little "far out" for you. If so, please research the ones that you have the most trouble with. When you do, I think you'll agree that they're worth trying. We've found that it is very empowering to catch a bug at the same time as a friend, be able to treat it ourselves while they go to a doctor and get a prescription, and feel better before they do.

Again, even if you intend to go directly to a doctor if you get the flu, I encourage you to study these alternatives and have some or all of them on hand. It could be very important, especially if there is a pandemic, people are dying, you get sick, and timely professional medical attention is not an option.

Chapter 5 Resource page:

www.urbansurvivalplan.com/193/lesson5

Chapter 6. Chemical and Biological Attacks and Ghetto Medicine

To begin with I want to address the 500 pound gorilla in the room, which is nuclear attack. I am not addressing nuclear attacks in this course for the simple reason that we're trying to address as many high probability disasters as possible in a limited amount of time. The chances of nuclear attack are several scales of magnitude lower than the chances of chem/bio accidents, conventional terrorist attacks, hurricanes/volcanoes/earthquakes, wildfires, or even economic collapse. At the same time, fallout shelters are expensive and take months to put in place.

I can point you to several resources to help you get prepared for nuclear events…and will do a follow-up lesson on the topic if the demand is high enough, but for now we're going to spend more time on events that are more likely to affect you and that have solutions that are broader in their possible applications.

How great is the risk of chemical or biological attack?

Biological warfare has been effectively used since the Hittites herded victims of the plague into enemy lands in 1500 B.C. and chemical warfare has been used in the form of poison tipped arrows since man began hunting. They are proven strategies that have benefited from thousands of years of use and refinement, but how big is the risk that we face today?

To begin with, let's look at Islamic terrorist groups that are operating openly in the US. Nineteen states have active, publically known terrorist cells ranging from the Muslim Brotherhood and Hamas to al Qaeda.

If you would like to see a map provided by Kim at The Investigative Project On Terrorism, please go to www.urbansurvivalplan.com/229/terrormap.

You can also find it in the resource page for this chapter at: www.urbansurvivalplan.com/306/lesson6resources

Second, the effects of chemical and biological weapons are truly terrifying. Blistering skin, suffocating, loss of nerve control, organ liquefaction, sores, and the thought of mummy-like infected people coughing and spreading suffering and death is enough to keep a sane person up at night.

Third, the Soviet Union manufactured hundreds of tons of biological weapons as well as 40,000 tons of chemical weapons. Some of these are still under Soviet control. Others were destroyed, sold to countries like Iran, and some of them and have been lost.

A friend of mine from Provo, Utah studied bio-weapons under the head of the former Soviet Bio-Weapons program, Ken Alibek and contributed a lot of valuable information for the previous lesson and this one. According to Mr. Alibek, the height of their production, they were manufacturing enough Anthrax to kill the entire world several times over, EVERY WEEK and had the capability to ramp up their production levels by several multiples.

Fourth, Both Iran and North Korea have chemical weapons stockpiles and are working on biological weapons.

Finally, we've seen both the manufacture and use of chemical and biological agents by individuals and terrorist groups in recent years, including:

- The use of sarin (chemical nerve agent) in the Japanese subway system in 1995. The attack happened in 5 subway cars at roughly the same time during morning rush hour. A single drop of sarin the size of a pin head can kill an adult. Each attacker carried at least 1800ml of sarin (almost 2 liters/32 ounces), most of which was released, but injuries were limited to 1100 and **12 died**.
- Post 9/11 anthrax attacks in 2001 by a former US government scientist, infecting 22, **5 of which died**.
- A foiled al Qaeda attack in Amman, Jordan in 2004. They had 20 tons of chemical agents, including 71 different lethal blister, nerve, and choking chemicals. The chemicals would be aerosolized and spread by three precise explosions designed not to burn up the chemicals. The combination of chemicals meant that victims would have multiple symptoms and there would be no single effective treatment. The effective kill zone of the chemical cloud was expected to extend outward 1 mile from the explosions.

- Three 2007 "dirty" chlorine attacks in/near Baghdad. A pickup loaded with explosives and chlorine cylinders was blown up, hospitalizing 55 and **killing 5**. In a separate incident, a bomb was detonated on a chlorine tanker affecting 150 local residents. And in the third incident, a pickup loaded with explosives crashed into a chlorine tank, **killing 12**.
- 2008 ricin event. A Salt Lake City man manufactured homemade ricin for "self defense" and mishandled it in a Las Vegas motel, causing him to be hospitalized.

Ironically, when you understand the full significance of these five sets of incidents, they are five of the most comforting things you can know about the risks we face from chemical and biological attacks.

To begin with, take a look back at the number of people killed in each attack. The deaths were all tragic, but they were relatively small compared to other attacks, like the bus suicide attacks that happen all too often in Israel using conventional bomb vests.

Next, take a look at the Japanese subway attacks. The loss of life was tragic, but the terrorists used enough sarin to kill tens of thousands (possibly hundreds of thousands) of people and instead, killed 12. The group responsible, Aum Shinrikyo, had millions of dollars at their disposal and had spent the previous 3-4 years trying to either buy or produce (with the help of several PhDs) either a toxic strain of botulism or anthrax. Despite having time, money, and skilled help, they were unsuccessful.

Then look at the proposed attack in Amman. The death toll projections ranged from 20,000-80,000 with well over 100,000 injuries expected. This is would have been a tragedy, but the silver lining here is that the number of people who would have been <u>within the 1 mile kill radius</u> of the chemical clouds and <u>outside of the kill radius</u> of the initial blasts would have been a limited number.

This is an important point, because it brings up the fact that even if there is a large scale chemical or biological attack, and the target is in the US, and the target happens to be in your city, the chances of you being directly affected are very slim.

If you aren't an immediate fatality in a chemical or biological attack, the biggest threat that you face will most likely be civil unrest as a result of panic, looting, and other civil unrest.

Finally, we have the 19 publically known terrorist attacks that have been thwarted since 9/11. It would make sense that if terrorists intended to use chemical or biological weapons, at least one of these attacks would be chemical or biological. While some of them planned on using conventional explosives to attack fuel targets or making a dirty-nuclear bomb, NONE of them included biological or chemical attacks, let alone using the nightmare scenario of using a crop duster to spread bio/chemical agents over a city.

Let's compare this threat to a few other threats:

1. Malaria, tuberculosis, & HIV/AIDS kill 5 MILLION people every year.
2. Smoking kills upwards of 5 million people every year (443,000 in the US.)
3. Ordinary flu kills an average of 36,000 people per year in the US.
4. The cumulative total from the chem/bio attacks mentioned above that span a 15 year period is 34, or roughly the number of people who die from the flu every 8 hours in the US.

As you begin digging into the threat of chemical and biological attacks, you'll soon read several nightmare scenarios that have been put forth by various local, state, and federal entities. At first, the effects of a chemical or biological attack look horrible, with large percentages of cities suffering and dying.

The statistics from these scenarios are used in news articles, fiction books, TV shows, and movies. They are also what most people think of when they think about the threat posed by chemical or biological attacks.

If you dig deeper, though, you start to see that the technologies used aren't realistic. In the case of bio attacks, Ken Alibek and his Soviet scientists were SCARED of making dry powder smallpox, because it is so dangerous that it's nearly impossible to work with without becoming infected, even when using the best containment and decontamination methods known. Even so, we "game" the use of dry powder smallpox in our scenarios and use infection and re-infection rates that are 3-5 times higher than what we know to be true or what scientists predict as a worst case.

With both biological and chemical attacks, the scenarios ignore how vulnerable potential toxic clouds are to environmental conditions such as temperature, humidity, elevation changes, wind (too strong or none at all), and micro weather patterns found in areas with lots of trees, hills, buildings, and even large black asphalt parking lots.

With both, the concentrations need to remain in a narrow range to achieve maximum death tolls. Too high of a concentration wastes the weapon and too low of a concentration turns many weapons from killers to severe irritants.

Fortunately, the response to a chemical or biological weapon attack is the same as the response to the chemical accidents that we addressed in Chapter 3, so even though the chance of (1) an attack happening that is (2) in your city that (3) affects you is miniscule, we can create a response that is simple, inexpensive, and effective enough to be practical. In addition, the exact same response will protect you short term in the event of a dirty nuclear attack.

Your response:

There are two responses you're going to want to consider if you have a chemical or biological event: Evacuate or Survive (Shelter) In Place.

If an event happens, the local government entity is going to suggest that everyone take actions that will make their response easiest and hopefully limit the number of casualties. This is not necessarily the best suggestion for YOU.

As an example, let's say that I'm home with my family this evening when I hear a massive explosion. I go outside and see a giant cloud of smoke going into the air off to the Northwest. I can't see the ground where the fire is, but I know that there's a refinery 2 miles in that direction. As I'm looking, I hear multiple secondary explosions and see that the wind is blowing the cloud towards my house with a light breeze.

I look at the trees in our neighborhood and estimate that the wind is blowing at 5-10 mph at ground level, giving us 12 minutes to get ready or leave, assuming that the wind isn't blowing harder at higher altitudes.

I know that the smoke cloud shouldn't affect us at this distance, but I'd rather we didn't breathe in the chemicals or be around as the chemicals/debris start cooling, clumping with moisture in the air and start falling back to earth. My initial hunch is that it's an accident or isolated attack, so I'm not worried about long term security issues.

This has all happened within 30 seconds of hearing the explosion, so there's nothing on the news yet and there's not likely to be any increase in traffic yet, so we grab our home GO bag, turn off the AC, lock up the house, top off our gas tank with our reserve fuel, hop in the car, and take off within 10 minutes of hearing the explosion.

As we're driving off in the car, we hear that there was an accident at the refinery and the "official" recommendation that people downwind from the refinery move indoors, seal their doors and windows and stay calm.

Many people panic and leave when they hear the announcement…taking 30 minutes or more gathering belongings and actually leaving. Gas stations have lines from people who left home with empty gas tanks and the roads leading away from the refinery are soon packed like rush hour.

Meanwhile, as the congestion is building, we're already checking into a hotel in a neighboring town that's perpendicular and well away from the path of the smoke and debris. Our simple knowledge and preparedness allowed us to respond quickly, decisively, and calmly and turn an event that was full of drama for others into little more than an impromptu night or two away from home.

What was the best response in this instance? There are several factors, but the speed that we were able to respond made leaving a viable option. If I'd ignored the sound of the initial explosion and tried to leave when the alert came on the air, we would have been stuck in traffic like everyone else and would have probably been better off Surviving In Place.

Another factor is the overall civil situation in your area. If you're in the middle of a larger emergency like being in New Orleans 4 days after Katrina, you might not have the option to leave quite so easily.

As I have mentioned before, if an event like this happens while your family is spread out between work, school, and activities, you're probably not going to be able to stomach the thought of leaving without them. I couldn't.

Survive In Place™

If evacuating is not an option, you've got two major strategies you can use:

1. For an airborne threat where the contents of the air is the threat.
2. For an epidemic/like the flu or another biological event where the air is not dangerous but other people are.

If you're facing an airborne threat, you want to try to seal off a room or section of your house where you can stay for at least a few hours, but possibly longer. Most chemicals dissipate within a few hours under normal atmospheric conditions, but an uncontained leak or an industrial fire could put contaminants in the air for much longer.

Ideally, you want to pick a room/area that can be shut off from the rest of your house, has no windows, no air ducts, and hard floors. If that's not a possibility, pick the room with the fewest windows and ducts. Try to allow for at least 10 square feet per person for adequate oxygen.

If you are in a situation where you need to use your safe room, make sure to take everyone's pulse every 10-15 minutes and write down the information. Your pulses will probably be elevated from stress and rushing to get into the room, but if you start to see a sudden spike in your pulse after the 5 hour mark, it could mean that you're running out of air and will need to prepare to leave your safe room immediately if anyone begins to pass out or is unable to stay awake. The 5 hour mark is merely a guide and will be very different for two professional free divers who have learned to control their breathing and heart rate than it will be for two large people who are prone to panic and hyperventilation.

You'll have to make a choice on what to do with your pets. They take air that you may need, you may have bathroom messes to contend with, and they may cut your plastic with their paws if they hear outside noises. We've made the decision to include our pets, but we only have two dogs and understand the risks. Now is the time to decide what you are going to do with your pets.

TO DO:

Decide what you're going to do with your pets in the event of an airborne threat.

Since many chemicals (chlorine, as an example) are heavier than air and will tend to collect in basements, you will ideally want to have your safe room on the ground floor or higher.

Your first step in securing your house is going to be to limit the amount of outside air that can get into your house by turning off your AC/furnace/fan, shutting and locking all doors and windows and closing your fireplace damper, if you have one. If you feel like you have time, tape a sheet of plastic over your fireplace opening.

Next, you want to create an interior safe room with an even higher level of safety. Here's how to create a safe room in your house.

First, you need the following items:

1. Your home GO bag
2. Thick plastic sheeting. You'll see recommendations on thickness ranging from "anything thicker than plastic wrap" to 10mil. We have 3.5mil. That being said, I wouldn't hesitate to use contractor trash bags or multiple layers of regular trash bags if I was away from home and that was all that was available.
3. 3" wide duct tape
4. 5 gallon bucket with a lid and trash bags for your "bathroom"/TP/wipes

If you've got time, you'll also want the following:

1. Pillows, bedding, blankets
2. Extra drinks
3. Extra food
4. Small books/games
5. A radio if you don't have one in your GO bag
6. A land line phone that's plugged into the wall and/or a cell phone
7. Chemical heat packs/foot warmers and/or chemical cold packs

As a note, everything except for the GO bag will store easily in a closet in your 5 gallon bucket.

Sealing the room

As a note, our safe area is a bedroom and bathroom "area" (I wouldn't call it a suite) that gives us water, a toilet, a bed, and makes our safe room quite comfortable.

Once everyone's in the room, the first thing you want to do is fill the crack under the door with trash bags, pillowcases, or clothes. Second, tape all the seams on your door(s), and window(s).

Third, using a utility knife, scissors, or a pocket knife, cut a sheet of plastic for each door, window, vent, electrical outlet, or other openings to the outside or the rest of your house. Make sure to allow an extra 3-5 inches of plastic on every side. When you're done, tape down the corners and then, using one single strip of tape per side, tape down the sides.

Thick carpet poses a unique threat in that you may not be able to get a good seal. Depending on the threat, you may need to find another room or cut through your carpet so that you can get a good seal between the plastic and the cement or wood below.

TO DO:

As a test and to prepare, buy a roll of 3 ½" **BLUE** painter's masking tape (it is designed not to damage paint), cut your plastic barriers, and tape them into place in your safe room with the BLUE tape. This will insure that you have your plastic pre-cut, that you know how much tape you'll need, and as with any skill, your second time will be much faster than the first.

Test out the blue tape on your walls before you go crazy with it. It's made to be easy to remove without damaging walls/ceilings/floors, but you will want to test it to make sure.
As an additional step, spend 3-5 hours in your safe room so that you have experience using your 5 gallon bucket and know if you need other items.

As a test to see how well your taping/sealing is, one strategy is to put yourself in the room while someone else cooks a pungent meal like curry, onions, or garlic in another part of the house. If you have a portable fan, blow it towards your safe room as an added test.

It should go without saying, but if you have thick carpet, do not cut it for the test.

At Work

If you are at work when you need to create your safe room, the basic procedure is the same, except you will have the extra challenge of blocking the air ducts.

The best solution is to find out how to shut off the air in the room you've chosen as a safe room ahead of time. If it's not obvious and you have maintenance personnel, they may or may not be willing to tell you how to turn off the air.

If you can't turn off the air, you can still block the air. Here's how:

1. If possible, find a room where the fan doesn't blow as strong as other rooms.
2. Get access to the duct by removing the grate.
3. Take an alcohol wipe or something similar and clean the duct near the grate so that your tape will stick.
4. Cut a sheet of plastic that is a couple inches bigger than the duct in every direction.
5. Tape the plastic sheet to the duct as close to the grate as possible. You may need to reinforce the tape by super gluing both the plastic and the tape to the walls of the duct. Depending on how strongly the fan is blowing, it may be very difficult to get a good seal.
6. Using a pillow, crumpled up paper, or anything else handy, fill the 3 ½ inch gap between the plastic and the grate due to the tape and re-close the grate. This will help support the plastic and take some of the pressure off of the duct tape.

In both home and work situations, once you're in your room, listen to your radio until you receive the "all-clear" to come out. You can also call or text your out-of-town emergency contact occasionally to see if they have any additional information about the situation in your area.

Epidemics and Pandemics

If your area is experiencing a biological event, be it the aftermath of an attack or if a highly contagious and nasty strain of the flu develops, you may want to isolate yourself from other people until the threat passes.

It's important to note that the limited number of epidemics and pandemics that the Earth has experienced means that there is not a pool of experts who have "been there" and "done that" to provide us with proven strategies to follow. Instead, I have consulted with survival experts and am sharing their combined wisdom with you. This section will be expanded and refined if there is a pandemic approaching and I will keep you informed of changes via email and blogs.

We're going to spend a few lessons on the subject of surviving without leaving your home, but there are a couple of items that are specific to biological events:

1. Do a "gut check" and make sure that things are really to the point where you need to isolate yourself from other people. Are you being overly paranoid? Is the risk of sickness great enough that it's worth risking your job and/or relationships?

2. Get your story straight. If you can't pull off looking and sounding like you're not home, put a sign on your front door explaining why you refuse to answer. It could be, "Wife pregnant, we're not answering our door" or "Baby sleeping, please don't knock or ring the doorbell" or "Mom just finished chemo, we're not answering the door" or something similar.

   I wouldn't suggest putting a sign up saying that you're sick as a ploy. The possibility exists that sick people will be gathered up so that they can be "helped" and isolated from the general population. It's doubtful that officials would believe that nobody is really sick in your house if you have a sign up saying that someone is.

   You want to try to create an impossible argument so that anyone reading the note will leave you alone and won't think you're just being rude. Whatever you decide on, you MUST stick with it and do not tell anyone that it is merely a ploy.

3. Decide what you're going to tell your employer, co-workers, friends, and family. They may not have prepared like you have and/or may not appreciate your concern or you isolating yourself from them. Try to be as nonchalant as possible and as non-committal as possible. If things don't get as bad as you think they might, you don't want to have burned any unnecessary bridges.

The most awkward time in a situation like this is going to be the time between when people who are "aware" start taking action and when the majority of the population realizes there is an issue.

4. Get daily sun exposure. If possible, get it where people can't see you and won't be tempted to talk with you.

5. Stay out of sight as much as possible, drawing curtains if necessary.

6. Decide your response if "help" comes.

7. Get daily exercise to help combat stress.

8. Decide early on what trigger you will use to start going back out into public.

9. Set limits on how much news you will watch, read, or listen to. The media's primary job is to keep your attention so they can sell advertising. It's not to motivate you, educate you, or help you through a disaster. I anticipate that the phrase, "If it bleeds, it leads" will be particularly accurate during a biological event.

10. Take OPSEC precautions when you are talking with people about the flu and your plans/preparations. If there is a break in the supply chain and they think that you are "hoarding" food, vitamins, or medication, they'll be coming to you when they need help. I'll paraphrase James Rawles by saying that charity is great during a crisis, but it should be anonymous and on your terms.

Ghetto Medicine

Otherwise known as "field expedient" or "improvised" medicine, here are some of my favorite non-conventional medical treatments. By reading the following and continuing, you accept the risk that these treatments may hurt you or others and should only be used after doing your own research, getting proper medical training, and as a last resort.

These tips, tricks, and shortcuts can help anyone, but are really designed to help people with formal medical training train their mind to see alternative methods of treating themselves and patients when they are in a non-ideal situation.

I am a former first aid and CPR instructor and believe that everyone should have this level of medical training, at a minimum. The classes will teach you how to respond to basic traumas and medical emergencies, as well as how to communicate with EMTs and Paramedics That being said, it will not teach you how to deal with injuries or medical situations when you don't have someone with more training coming to relieve you quickly.

It's for that reason that I encourage you to get wilderness medical training of some sort, like wilderness first aid, wilderness first responder (WooFeR), wilderness EMT, or Outdoor Emergency Care (Primarily for ski patrol, trail patrol, and search and rescue). Why? Because all of the wilderness medical training classes assume that you're going to be away from an ambulance, helicopter, and other higher levels of medical care and supplies for an extended period of time. These classes force you to improvise, adapt, and overcome to treat people with what you happen to be carrying with you and what you can find in your environment. This skillset most closely fits what you'll be facing in a disaster or emergency situation.

Here are some of the tricks that either I or one or more of my contributing experts, including a former Navy Seal Medic, Combat Medics back from Iraq and Afghanistan, local paramedics, and first responders who operated in New Orleans after Katrina have used over the years that you will surely benefit from.

1. **Superglue/Crazy Glue.** At the time of this writing, I have used superglue or crazy glue on 4 facial cuts, a cut on my knuckle that went to the bone (I was on a hunting trip and was fortunate not to have cut any nerves, tendons, or ligaments), and numerous smaller cuts on myself, my dog, and others, as well as blisters and hot spots.

   I've found that Crazy Glue works a little better for me (dries quicker and thicker), but the 12 packs of superglue are so convenient that it is the main brand that I use. It's cheap enough and has enough uses that it's worth buying a few tubes and practicing with it on your hands and/or feet. Superglue is very effective at securing skin to skin, so do not touch the glue with a finger to see if it is dry. Instead, use a piece of string, floss, paper, or a similar item that you don't mind being stuck to you for a few days.

As a rule of thumb, if I am on the fence about whether or not I need to get stitches, I use superglue. If I were to get a serious cut where the muscle needed to be sewn together, I would not use superglue. In "normal" times, I'd get medical attention and in a disaster, I would attempt to use a suture kit or dental floss and a needle if I absolutely had to.

Some people have an allergic reaction or asthmatic reaction to Superglue. Don't wait until there is an emergency to find out if that's you. Take the time to try a small amount of it on your skin and see if you have a reaction. Make sure you do this in a well ventilated area and have a responsible adult with you who can get you to medical treatment if necessary when you do this.

As a note on personal responsibility, allergic reactions and severe asthma can kill you. You are responsible for your actions. DO NOT try this if you have ever had an allergic reaction to superglue, anticipate that you might, or if you are apprehensive about trying it.

If you aren't allergic to it and decide to use Superglue, the first thing you want to do is irrigate/clean the wound and control bleeding with pressure and elevation. Next, clean it out with sterile water or saline and pat dry. Then, close the cut with a butterfly bandage or tape that you've cut like a butterfly bandage.

When the skin is lined up the way you want it to heal, apply a thin layer of superglue. It will sting, and burn both your cut and possibly your eyes. If the cut is 1" long, you will want to apply the superglue approximately ¼" to ½" to the side of the cut so it has enough healthy skin to grab onto. Spread the superglue with a toothpick, metal pin, or the tip of the tube if you must, but never use a Q-tip or cotton. Once the glue dries, (in 1-2 minutes) apply a second layer over the top.

As a final step, cover the superglue when it's dry with athletic tape, duct tape, or a bandage making sure to use gauze, cloth, or paper to keep the tape from actually touching the superglue. The butterfly/superglue will hold the wound closed and the covering will help the superglue stay in place as long as possible and keep the wound area from getting dirty.

You can also use superglue on hot spots and blisters, but I've found duct tape to be superior in almost every way for this application.

Make sure to use superglue in a well-ventilated area and try to avoid breathing in the fumes or letting them hit your eyes, especially if you wear contacts. Inhaling the fumes can cause flu like symptoms or an asthmatic response. It can also cause an allergic skin reaction, so make sure to test it in a non-emergency situation first. You also want to avoid touching superglue with cotton clothing, cotton balls, or Q-tips, as it causes an exothermic reaction that can burn or irritate your skin.

There are less toxic "medical" superglues available including several brands of liquid bandage and even prescription strength Dermabond. I've tried every commercially available liquid bandage I've seen and I haven't been impressed with any of them. They are less toxic, but they just don't work that well. In my experience, they dry slower than superglue, wear off faster, and don't hold wounds together as well.

The BEST liquid alternative to stitches that I've used has been Dermabond. Unfortunately, it is $20-$50 per single use tube compared to $5 for a 12-pack of superglue that gives you the possibility of multiple uses per tube. It is the perfect combination of fast drying, high strength, high durability, and low toxicity. I had an ER doctor use it on a cut directly under my eye with minimal effects from the fumes.

Because of the toxic nature of superglue and the reactions that it causes in some people, I can't categorically tell you that it will work as well for you as it has for myself and others. As a result, I encourage you to try out some of the "medical" superglues that you can find anywhere first aid supplies are sold.

2. **Maxi Pads.** There are some medical/hygiene items that there are few good substitutes for. Maxi pads are one of those items. If you live with or spend time with any women who are in their menstruating years, it's just smart to have maxi pads in your 72 hour kits and/or med kits.

   I carry maxi pads for another reason, and that is to use as absorbent trauma dressings. They will soak up a lot more blood than a 4x4 and make more sense than ruining a roller bandage simply soaking up blood and applying pressure.

A note on tampons. I used to carry tampons to pack puncture wounds from sticks, poles, and presumably from bullets and stab wounds. Over the last few years, EMT and medic experience in the field has shown this to be a bad solution for small injuries because the tampon swells and tears healthy surrounding tissue as it soaks up blood and expands. You'll still see people promoting the use of tampons and duct tape, but I strongly suggest that you do NOT pack wounds with anything that will become an integral part of a clot and has to eventually be removed. If you get the patient to a higher level of care, tampons are a pain to remove once clotting has started and if you are looking at leaving the wound packed for an extended period, you risk septic infection.

If you're a guy and you're not up to speed on feminine hygiene products, tampons are cotton "plugs" with a string attached for easy removal and maxi pads are rectangular absorbent pads with an outside coating that helps to keep the clotting action of the wound from attaching the pad to the patient.

Two other alternatives to absorbent dressings that you may have with you in an emergency are disposable diapers and adult undergarments.

3. **Dental Floss.** I learned about the wonders of dental floss when I was doing ski patrol. One of the issues that you face with wrist, elbow, shoulder, or arm injuries is swelling of the fingers. If you are wearing a ring and your finger swells up to where you can't remove your ring, you have a potentially serious problem. If the finger keeps swelling, the ring will eventually act like a tourniquet, so it's important to remove the ring as soon as possible.

   Unfortunately, you aren't always able to start providing care before the finger is too swollen to remove the ring. If you do research on ring removal, you'll find people giving advice about using every type of lubricant known to man, but dental floss is the simple solution that is widely regarded as the best by EMTs and ER nurses.

Try to get the widest waxed dental floss you can find and start wrapping the ring finger, starting at the fingertip. Wrap the floss around the finger all the way up to the ring, making sure that you wrap tight enough and close enough together that the finger + floss is smaller than the ring. Once you get up to the ring, do a few more quick wraps to push down sections that you've missed and remove the ring quickly.

As a note, you can also use VCR or cassette tape for this.

TO DO:

Try this on yourself. One of the things that you'll notice is that speed is key with this technique. Your finger will start to swell as you're wrapping it and you will have to go back and wrap over sections that have swollen in the few seconds since you wrapped them. This effect will be even more pronounced with someone who is experiencing distal swelling due to an injury and speed will be even more important.

4. **Sugar.** In the movie "Shooter", Mark Wahlberg poured granulated sugar into his bullet wound to help it heal. Believe it or not, this treatment has been WRITTEN about since 1700 BC, is known commercially as "Sugardine" or "Sugardyne" and is a very popular treatment for injuries in horses.

   The simplest and most effective mixture is sugar and a 10% iodine or betadine solution. Simply pack or cover the wound with the mixture, and dress it.

   The mixture works by in part because bacteria don't grow in the presence of pure sugar, the sugar absorbs fluids, promoting drainage, and the betadine/iodine prevents infections.

   Even after 3700 years of use, the instructions for how often to re-apply vary widely from every 4-6 hours to every day. The common advice is to apply a fresh mixture when the old mixture has completely liquefied.

5. **Gunpowder**. This technique was made popular by John Rambo when he packed a wound with gunpowder and lit it to cauterize the wound. **Please don't ever attempt to cauterize a

wound with gunpowder.** The basic premise of using gun powder on a wound actually has a basis in truth.

Gunpowder, otherwise known as black powder, is a combination of sulfur, charcoal, and saltpeter. These were common wound treatments for over a century and are still generally accepted naturopathic treatments for wounds.

Unfortunately, the modern handgun and rifle ammunition that you shoot does not contain gunpowder. It contains smokeless powder, which is usually nitrocellulose or a stable form of nitroglycerin. In any case, modern ammunition does not have sulfur, charcoal, or saltpeter and will burn like heck and not do much good in a wound.

6. **Safety Pins**. Safety pins are wonderful survival tools. In addition to using them as an improvised fish/animal hook, zipper handle, and a quick fix for shoe laces, they have several medical uses as well. You can use large safety pins for finger splints, to drain a blister, close large wounds to prevent debris from entering, and even as an improvised sling.

   To use safety pins as an improvised sling, all you need is an injured patient with a long sleeve shirt on. Simply put their arm in the position that hurts the least and start pinning the sleeve to the body of the shirt. Once you're done, they'll probably still be able to move their arm around because of the shirt being loose against the body. To fix this, simply draw up the extra material under their good arm and knot it or pin it so their shirt is tight against their skin and the injured arm is immobilized as much as possible.

7. **Duct Tape**. Duct tape is the survivalist's friend. Here are 16 more reasons to carry at least a full roll of duct tape in your kits.
   a. Hotspots (pre-blisters) – apply directly to the area, making sure to cover an area bigger than the hotspot.
   b. Blisters – cut a piece bigger than the affected area, put paper/cloth between the blister and the tape and you're good to go. Depending on the blister, you may/may not need to build up the area around the blister to

relieve pressure or pop it to continue using the affected area.

c. As a temporary cast, once you've set the bone, you can wrap the bone/joint in newspaper and then wrap the newspaper with duct tape until you get to a higher level of care.
d. If you are wearing dress shoes or cheap shoes that are falling apart in a survival situation, a few wraps of duct tape will get you more mileage out of them.
e. If your shoes are completely thrashed, you can make improvised shoes from duct tape by wrapping your feet. Place extra layers in front of your toes to give them a little extra protection from stubbing. To pad the bottom, you can use card board, carpet strips, or anything else you find in your environment.
f. Gaiters. Since blisters are so detrimental to effective movement, if you find that you're repeatedly removing pebbles/debris from your shoes, you can make duct tape gaiters by wrapping tape around the top of your shoes & legs a few times to make sure no debris can get in your shoe.
g. Knee support. Many people wear a tight wrap under their knee to provide support. Wrapping a ¾" strip of duct tape around the leg twice can accomplish the same thing.
h. Knee support. If you experience a torn ACL/MCL and don't have knee support, you can splint your knee with duct tape and any available stiff object (stick, bar, rolled newspaper, file folders, cardboard, etc.).
i. Sutures/stitches. If you can hold a wound together temporarily while you apply duct tape (with superglue butterfly bandages, or improvised butterfly bandages) you can provide additional lateral support to the wound to keep it from re-opening. Just remember to put something (paper, cloth, gauze, etc.) between the wound and the duct tape so the duct tape won't stick to the wound and open it when you remove the tape.

j. Sucking chest wound. Simply tape 3 sides of a plastic bag over a sucking chest wound to create a butterfly valve.
k. Neck brace substitute. If you've got someone secured to a board, chair, etc. you can immobilize their head with duct tape instead of the traditional wrap. Just make sure to put something between the sticky tape and the patient's hair.
l. Dog paw protection. If you're in an area with sharp rocks, ice, or debris, you can tape your dog's paws to protect them. They will lose significant traction, but saving their pads is often worth the tradeoff. You can wrap them initially with a roller bandage to keep the tape from pulling out their hair when you remove it.
m. Finger splint. Use duct tape along with a popsicle stick, safety pins, a Bic pin, a folded file folder, or even rolled up duct tape!
n. For mass casualty incidents, you can use duct tape stuck to the patient to record triage information, as well as vital information.
o. As an alternative to triangle bandages for providing an anchor point to apply traction to arms or legs.
p. Arm splint. Find where the arm is most comfortable and secure it to the body.

TO DO:

Get a roll of duct tape and try several of these techniques on yourself and with a friend or family member. Critique each attempt and try to figure out why certain methods work better than others. Be very careful with duct tape and body hair so that you don't end up looking like a man-o-lantern. You may want to shave a small area or do most of the exercises over clothes to minimize pain.

Review of This Chapter's Assignments

To Do:	Date First Completed:
Decide what you're going to do with your pets in the event of an airborne threat.	
Do a test run in your safe room using BLUE painters tape. Monitor everyone's pulse to get used to being able to take vitals quickly.	
If you know that you aren't allergic and have no medical concerns try using superglue on your skin to see how well it works for you and how long it stays on your skin.	
Practice using dental floss to remove a ring from your finger and from someone else's finger.	
Practice some of the duct tape medical treatments.	
Make sure you have duct tape, pads, superglue, safety pins, and wide floss in your 72 hour kits.	

Chapter 6 Resource page:
www.urbansurvivalplan.com/306/lesson6resources

For more information on entering/exiting safe rooms go to:
www.urbansurvivalplan.com/299/saferoom/

Chapter 7. Building Your Own Team – AKA Mutual Aid

Much of survival, like life, is about relationships. It's great to have lone-survivor skills, but for most people, it is more practical to survive disasters with someone else. I've gone 6 chapters without mentioning much from the Bible, but there is a quick section from Ecclesiastes 4, taken from the King James Version of the Bible that is very applicable for this chapter:

9 Two are better than one; because they have a good reward for their labor.

10 For if they fall, the one will lift up his fellow: but woe to him that is alone when he falls; for he doesn't have anyone to help him up.

11 Again, if two lie together, then they have heat: but how can one be warm alone?

12 And if one prevail against him, two shall withstand him; and a threefold cord is not quickly broken.

If you are ever in a situation where civil order has broken down you are most likely going to want to have a put team in place ahead of time that you know you can depend on. Joining forces with other people will allow you to "switch off" your mind and get good sleep, recuperate faster if you are injured or sick, split laborious tasks, and buoy each other's spirits. In addition, joining forces will allow you to benefit from other people's skills, experiences, and resources.

Again, this doesn't just need to be an extreme disaster, like a terrorist attack or a full breakdown of civil order. Having a team in place is valuable for everyday disasters like floods, wildfires, hurricanes, and tornadoes.

When I've explained this concept to people before, one of the first responses is, "It sounds like you're showing people how to make a mini-militia." This is an inaccurate use of terms. Your group should be willing to help protect each other from physical attack, but most of the benefits of a small survival group are much more mundane. A better way to look at your "team" is as a mutual-aid group.

Let's look at a quick example of this. Many "aware" and "switched-on" people in the US happen to be in their 60s and 70s. They are in the prime of their life as far as knowledge and wisdom and many have their survival resources in order from years of making frequent, small purchases. Unfortunately, they don't have the stamina or strength that they did when they were younger, and may even have frustrating physical disabilities.

There are others taking this course who are in their mid 20s. They are full of energy, willing to take reasonable risks, and may even have great skills from Scouts, growing up hunting, or from law enforcement/military service. In any case, they don't have the financial resources that their older counterparts have.

If the personalities match, and expectations were clearly defined, a couple in their 60s teaming up with a single person or young couple in their 20s could make a great match. They both have strengths and shortcomings and can both benefit tremendously from the relationship.

The older couple can share wisdom from decades of living as well as resources that the young couple would have to acquire over a course of several years. The young couple can do physical tasks much easier than the older couple and are a bigger deterrent to potential looters. In addition, if they are in a situation where they need to do a night watch, they can split it into 4 shifts, rather than only two.

If you remember the "Skills Assessment" that you completed in Chapter 4, you identified areas where you had strengths and weaknesses. You also identified the people who you currently know who live near you who are the MOST skilled in each particular area. We're going to be using that information today to help pick out possible members for your team.

So how do you go about putting your team together? That's kind of like asking someone, "So, how do I meet a spouse?" There are as many ways to accomplish this as there are people reading this, but I'm going to start off by telling you some things NOT to do:

1. Don't tell your neighbors, "I'm ready for anything…I've got a year of food, medical supplies, and everything I need to stay safe if there's a disaster."

2. Don't open a conversation by saying, "I've got lots of survival stuff…you want to watch each other's back if there's ever a problem?"

3. If you've got survival food/supplies in your garage, don't leave them in the open where people can see them when your garage door is open. cover them up or mark them as something else, like "cross stitch from Mom"

4. Don't approach someone about teaming up simply because they have a lot of guns or martial arts experience.

5. Don't drone on to your friends/family/co-workers/etc. about preparedness and how important it is for them to start catching up.

6. Don't team up with someone who has announced their preparation plans to everyone who will listen ever since Y2K.

7. Don't try to approach the subject by using cryptic comments like, "If the zombies start knocking on my door, I've got everything in place to take care of them…if you know what I mean." <<This is an actual comment I heard in a gun store.

I know people make these mistakes because I was making almost all of them before I started getting my plan picked apart by experts, and I still hear people making these mistakes almost daily.

In every case, these strategies expose you to unnecessary risks. We talked about this early on when we discussed Operational Security, but it's worth mentioning again. The risks that you assume when you talk openly about your preparations increase both your current risk of being burglarized and your possible risk of being looted/attacked in the event of a disaster.

Keep in mind that laws and laws of common decency go out the window when you are dealing with the parent of a hungry child. When you ask most parents how far they would go to feed their child if they hadn't eaten in days and were starving to death, they would say that they would do whatever they needed to in order to protect their child.

If they KNOW that you have a shelf full of #10 cans of food, how understanding do you think they'll be if you turn them away? If you give them food, who do you think they'll come back to in a day or two when they're hungry again? Do you think they'll come back alone or bring their friends too?

You see, the problem isn't helping one family, one time.

The problem is becoming another family's primary source of nourishment when you only have supplies for your own family.

If they bring their hungry friends, you may find yourself helping a dozen families repeatedly when you only prepared to take care of you and your family and possibly some charity.

Some of the stories that have come out of New Orleans after Katrina make this point painfully clear. People who were known to have supplies on hand and people who openly helped friends and neighbors early on often faced an unwelcome reality when they started turning away people.

Once they were identified as a source of food, hungry neighbors felt that they were entitled to that food. Some of the intimidation and revenge techniques included throwing rocks/bricks at houses, attacks on pets, kidnapping and **holding pets hostage in exchange for food**, assault, and (supposedly) even gunfire.

Add to this the reality that many people's survival plan consists of buying enough ammo to take whatever food they need by force. Just two weeks before writing this, I stood in a circle with eight active duty military talking about how quickly things were crumbling. When the conversation moved to preparations, I found myself listening to them mock families who stockpile food and detail their plans on how to "liberate" it if they ever need to.

Some of the comments were, "If they're dumb enough to buy food instead of guns, they don't deserve to live." and "When it comes down to it, my family is more important than their family and I'll take what I need." and "I don't have room to store a year of food, but I do have room for another case of .223."

I'm guessing that some of the guys in that circle were like me and just played along so as to not put a target on our backs, but there were enough active participants in the conversation for it to be unsettling. I love our military, and this plan is not pervasive among military personnel, but it is all too common among the general public. Unfortunately, I overhear this exact line of reasoning at least 1-2 times a month as I'm in stores, restaurants, and at gun ranges.

It's for this reason that I want to share a "mature" analogy with you. While it is not a perfect analogy, there are a lot of similarities between having sex and information sharing. In both cases, if you get intimate too soon you can get taken advantage of. Neither of them can be "undone", and the consequences of doing either with the wrong person can be long lasting.

You've probably made the mistake of sharing too much information about your survival plan already. Since you can't change the past, the only thing you can do is start moving forward smartly. We'll be covering a strategy of "decoy caches" in a couple of chapters that will help limit the damage.

Another analogy (I must admit that it is a little overly dramatic) is that building up your team while keeping operational security is like cultivating a spy network in a hostile country. In both cases, you want to get as much information as possible without divulging any more than necessary and ideally, you want your level of preparations to remain unknown to everyone except your key people.

The simplest way to look at team building is to treat it somewhat like you would dating and trying to find a spouse or a business owner trying to find a key employee.

Whichever lens you decide to look at the team building process through, there is a lot of overlap between finding a spouse, key employee, and building a spy network, so don't feel like you need to get too tied to any particular one.

There are several ways to find members for your team, but here's a GENERAL 7 point plan:

1. **Determine Your Team Profile:** Figure out what kind of person/people you're looking for. Use your skills assessment from earlier to help clarify what you need.

2. **Identify Possible Individuals/Families:** Start identifying people who you think would fit. (Remember in Chapter 4 when I told you that you were going to use your Skills Assessment in a future chapter? Now is the time.)

3. **Make First Impressions:** Get to know the person on a surface level.

4. **See If You're Compatible:** If the initial conversations go well, start asking them easy questions about politics, the economy, self-reliance, disasters, etc.

Make them conversational and comfortable so that it doesn't feel like an interview or an interrogation.

Not everyone will be compatible. Think of this like fishing. Sometimes they bite, sometimes they don't. Either way, just keep on casting.

There are articles in the paper and spots on TV almost every day on survival, so you can ask what they think about the article or TV spot without revealing anything. Try to ASK as many questions as possible without "showing your cards" (include spouses if it's a married person.)

Don't force a relationship that's doomed from the start. If they aren't on the same page as you, move on.

5. **Test the Relationship:** Try to do some activities together that will let you see each other under LIGHT stress. This could be a household project with a tight deadline (working with concrete is a good one,) primitive camping, or even a road trip or car camping.

   It could also be as simple as going skydiving or bungee jumping. It could also be taking a class together that neither of you are skilled in. In any case, you want to see how each other responds when you're outside of your comfort zones.

   Test the RELATIONSHIP, not the PERSON. The last thing you want to do is make someone feel like you're trying to make them fail.

6. **Test the Waters:** If everything goes well up to this point, tell them that you're interested in finding a few people who are willing to get the supplies and training necessary to survive a local or national "incident" and commit to helping others in the group regardless of whether or not an "incident" ever happens.

7. **Move Forward:** Develop group and individual goals for supplies, skills, and advanced training together and start sharing information. Make it clear from the onset that operational security is key and that everybody is expected to have supplies above the minimums that they do not tell others in the group about.

Who are you looking for? What kind of attributes are you looking for in your group? It's going to depend from individual to individual, but the more attributes you have on your list, the better group you'll end up with BUT it also means you'll have to go through a lot more people to find your group. Here are a few key things to look for:

1. Trustworthy
2. Similar or complementary worldview (religion, politics, self reliance, etc.)
3. Tolerance for different habits, opinions, and personalities
4. The closer you live to each other, the better. If you're next door neighbors or live across the street from each other, it will allow you to literally watch out for each other, give each of you a retreat location if something happens to your house, and it will allow you to have a larger group and still have some space.
5. Mentally resilient
6. Adventuresome
7. Adaptable
8. Willing to work
9. Willing to learn
10. Someone you'd trust to raise your kids.

You're going to have your own attributes that you consider key. Please share them with myself and the other students by posting them on the blog for this chapter at www.urbansurvivalplan.com/320/lesson7t

How big of a group?

I suggest that you decide this before starting your search for other team members. You're going to want a small number so that disease spread and OPSEC aren't problems, but a large enough group so that you'll have a wide range of skills and enough people to have one group venture out while another protects your home base. Keep in mind that if you don't have neighbors on your team, or a very large house, you'll be limited by the size of your house and the duration of the incident.

If you have the space, I agree with James Rawles assessment that 10-12 people is the ideal number. It is a large enough group to allow for people to have down time due to sickness/injury without being a logistical nightmare. Most people are not going to be able to fit this many people, food, water, and supplies into their house, condo, or apartment, so you will have to adapt the number to fit your individual situation.

A few days into an incident, you'll probably have a good idea about who in your neighborhood is "switched on" and who is in trouble. (try to appear as if you're barely making it) Once these "switched on" people start showing themselves, you can consider joining up with them, but remember to always underplay your resources and skills. You don't want to become the neighborhood convenience store or source of free labor unless you KNOW you have more than enough of both to give.

Strategies for finding people.

1. Start with people you know and like.

2. Promote SurviveInPlace™ locally. This is a shameless pitch, but it is also good, solid advice. If you promote the course through Clickbank, you'll get 50% of every sale (and rebill) after Clickbank's fees.

   So if a few people sign up through you, this book will be paid for and you'll be making money. If you are going to do this, let us know your Clickbank ID and where you live and we'll create a special URL (SurviveInPlace.com/Dallas or something similar) that is simple for you to send people to. Just send us an email to SIPBook@surviveinplace.com

   As people sign up, you'll get their email address (but none of their billing information or their address) and you can contact them and start an anonymous conversation with them by email. Here are some free/cheap strategies for promoting it:

   a. Flyers at coffee shops with the URL on tear away tabs

   b. Flyers at gun ranges

   c. Craigslist.com

   d. Survivalism.meetup.com

e. Backpage.com

f. Kijiji.com

If you want a sample flyer, let me know and I'll post it on the forum.

3. Announce yourself (using a handle) in the meet up section of the SurviveInPlace™ forum

4. Take a local CERT (Civilian Emergency Response Team), search and rescue, or a local multi-day firearms class. In each case, listen a LOT and share very little specific information, even if asked directly. Arrive early so you can sit in the back of the room and watch the other students.

5. Your city/county may have a formal mutual aid or sheriff's auxiliary group in place to help law enforcement and first responders during emergencies.

6. Local John Galt or Sarah Connor groups.

7. 2600.com or lock picking groups. << OPSEC is VERY important in these groups, but I've found their social engineering/hacking habits to be valuable.

8. Tea party or "5000 Year Leap" groups.

9. Join an existing group. Groups of people agreeing to help each other through thick and thin are not new and many formal and informal groups already exist. You can find existing groups looking for additional members on survival.meetup.com and you may find less formal groups in your own area during your search.

They may be a group of guys from your local National Guard unit, local firemen/police/medical professionals, or other similar groups.

If they let you into their circle, military and civil service units could be a great fit for your family. In the event of a disaster, they will most likely be hot bunking at their station, if they are resting at all and won't be able to watch over their family. If they realize this and accept you, it can make you a welcome member of their group and give you access to resources that you may not be able to get otherwise.

Asking and answering questions

With any of these strategies, a good segue into the topic (Step 4, seeing if you're compatible, above) is something like the following, "I've been thinking about taking a course I saw online called, "Survive In Place™ ." It basically walks you through the steps to survive a disaster in an urban environment if you can't leave. With all of the stuff I'm seeing in the news, it seems like I need to do SOMETHING to get ready. It sounds a little crazy to be saying, but…well…I don't know. What do you think?"

Or

"I saw a special on Hannity the other night on urban survival and it got me thinking. With all of the stuff I'm seeing in the news, it seems like I need to do SOMETHING to get ready. It sounds a little crazy to be saying, but…well…I don't know. What do you think?"

These are inviting, open ended questions that make people feel free to throw out their opinion without risk of being judged.

At some point after asking this, you're going to be asked questions. Here is a strategy that you can use early on in these relationships when you're posed with questions.

If they ask if you've done anything to prepare yet, you can say something like, "we've got a few cans of rice and beans, but nothing major." This makes it sound like you don't have very much at all and the terms are so relative that you are not lying.

If you are asked directly how much you have, you can also say, "I know we've got at least 5 cans…maybe 10 or so." This will be an honest answer, regardless of whether you have 10 or 100 cans.

I like this strategy of admitting to having a small amount of supplies because it allows me to tell the "truth" without telling the whole truth unnecessarily.

Rough waters ahead…

Accept the fact right now that you will probably screw up a few times along the way in forming your team. Did you ever date someone who you didn't marry? Did you ever hire someone who didn't work out as well as you thought they would? Unfortunately, that's reality.

Keep in mind that, even though 95% of marriages are "true love", 50% still end in divorce. There are no statistically significant numbers on "mutual-aid" groups like what we're discussing in this chapter, but it's fair to assume that the numbers will be similar to dating/marriage numbers.

With these facts in mind, there are a few common practices that you might want to avoid:

1. Don't co-mingle supplies unless they are clearly separated and marked. If your neighbor has a lot of storage space and you go in on a bulk buy together, spend the extra $10-$20 to get big Rubbermaid storage containers where you can store your portion.

2. Don't buy property, vehicles, or other items together. Have each person/couple buy things separately. Let's say that you and another couple both want a roto-tiller and a Big Berkey water filter and they happen to be about the same price. Instead of pooling your money and both buying both of them, simply have one couple buy the roto-tiller and the other buy the Berkey and agree to share them with each other.

3. There is no reason to share all of the specifics on what supplies you have on hand.

Unless you've worked in an undercover role at some point, you're skills are not going to be perfect. I encourage you to not hold yourself to a standard of perfection that is beyond your training. At the same time, take every conversation as an opportunity to improve your operational security skills.

Have fun with your group.

Don't make your group JUST about surviving the end of the world. Make sure to remember to enjoy life together. Have dinners together, do projects together, take classes together. If you've got 3-4 couples, you could even bring in a marriage counselor from your church to lead a couples study on relationships for a few weeks. Get creative. If you already have a group and have some fun, practical stories or ideas for activities to do together, please let us know at the resource page for this chapter www.urbansurvivalplan.com/320/lesson7t

This Chapter's Assignments

To Do:	Date First Completed:
Go over your skills assessment from Chapter 4. Update if necessary	
Practice your "half-truth" responses to questions about your level of preparedness.	
Start "casting" or talking with people about survival news that you see or read about.	
Start writing down what you want your group to look like…number of people, proximity, common traits, etc. Share them at www.UrbanSurvivalPlan.com/320/lesson7t	
Make a post on the forum in the "Local Meet Ups" section.	
Check local CERT class schedules at https://www.citizencorps.gov/cert/	
Look into local search and rescue, auxiliary, and mutual-aid programs.	
Sign up for a Clickbank account so you can get paid to promote SurviveInPlace.com > www.surviveinplace.com/clickbank	
When you've got your Clickbank ID, put it into this link and start promoting on Kijiji, Craigslist, and Backpage. www.YOURCLICKBANKID.urbsurv.hop.clickbank.net	

Chapter 7 Resource page:

www.urbansurvivalplan.com/320/lesson7t

Your SurviveInPlace™ "Gut Check"!

One of the big points that I made in introduction is about the importance of taking bite-sized chunks of information and actually taking action with them. The reason is this:

You Will ONLY Remember 10% of what you read but you'll remember 90% of what you read, hear, say, and ***DO.***

There aren't solid statistics on how these numbers change under stress, but from my own experiences and the experiences of the military, law enforcement, and survival experts that I've consulted with, it's safe to assume that you can cut these numbers in half again (at least) when you find yourself in a survival situation.

The lack of sleep, increased stress, and the need to make multiple critical decisions quickly combine to make some of the simplest tasks difficult. I remember the first real patient that I assessed after I became an OEC Technician (Wilderness EMT.)

For the life of me, I couldn't remember whether to take the pulse for 6 seconds and multiply by 10, for 10 seconds and multiply by 6, or 15 seconds and multiply by 4. It was ridiculous on several levels. I'd taken hundreds of pulses, but when I was working on a real patient, my adrenaline started racing and simple things weren't so simple anymore.

I want to do everything that I can to make sure that you leave this course with confidence **EARNED** from having a written plan that you can refer to, solid logistics, and memories of exercises you've done that you can tap into in a survival situation.

It's for that reason that we're not going to cover any new materials in this chapter. Don't worry; you're still going to get all of your lessons. This chapter is a "gut check," and could very well be more important than any other lesson.

Why?

Because, if it gets you to take action and turn your "head knowledge" into a written plan and/or experience, you will have made valuable progress in your urban survival planning.

Go through the checklists below and evaluate whether you have just been reading the course, or if you have actually been taking action. For the items that you haven't completed, give them a rating (1=I can do it RIGHT NOW and 10= difficult, expensive, or takes too much time right now.)

Using your ratings, try to complete as many of the preparations as possible this week.

Lesson 1: Operational Security, Scenarios, Improvised weapons

To Do:	Date First Completed:
Come up with a toned down answer for when people ask you what preparations you have made for disaster. Practice saying it until it is natural. Write it down in your SurviveInPlace™ Plan.	
Walk all through your house with the eyes of a stranger and see if your survival preparations are obvious. If they are, either camouflage your supplies immediately or make a written note in your SurviveInPlace™ Plan to make the necessary changes.	
Go through as many scenarios with yourself, your spouse, and your children as they are willing to do every day this week. (Max 2-3 per day with your spouse and children, unless they ASK to do more).	
As you go through these scenarios and break them down, make a note of them on 3x5 cards and keep them with your SurviveInPlace™ Plan with the scenario on one side and the solution on the other. Review and update them as necessary until your family has internalized them. If you don't have 3x5 cards yet, just tear a piece of printer paper into 4 pieces and use that.	
Identify at least 3 items that you could use as a weapon in every room in your house.	
Buy a detailed map of your city. The first map you should buy is a foldable map. In addition, you can buy a "book-style" map, but it is not strictly necessary.	

Lesson 2: Disaster basics

To Do:	Date First Completed:
Complete the following sentences for as many items as you honestly can: "I would walk through fire for…." "I would walk through fire to…." Put the written list in your SurviveInPlace Plan.	
Write/type your important information and contact numbers on a business card. Start carrying it with you and review it often so that you will eventually memorize it. Print it out on water resistant stock if possible (Avery 8878) but don't wait to take this step. Use an old business card of yours or someone else's and simply write on the back of it until you have business card paper to use.	
Create your prioritized list of Survival items to buy or trade for. Put the written list in your SurviveInPlace Plan.	
Make sure you can turn off your utilities	
Add water bottles to your freezer.	
Spend an evening/night with the utilities turned off. Write down what worked well and what issues you need to figure out before spending 72 hours without utilities.	
Start writing down potential dangers and congestion areas for your area.	

Lesson 3: Local threats

To Do:	Date First Completed:
Update your map with threats and choke points near and between your house and place(s) of work/school/relatives.	
If your home/work is vulnerable to a large scale chemical release, decide on a response.	
Assess your Urban Survival Skills/Weaknesses	

Lesson 4: 72 hour kits / GO bags

Go through lesson 4 to figure out what items to get for your kits and see how much more you need to buy. Make sure the bags are in place and you know how to use everything.

Which Bag?	Wish List Complete	All Items Bought	Bag In Place	You know how to use everything
EX: F-150	1/13	2/1	2/4	Yes
Home				
Vehicle 1				
Vehicle 2?				
Work?				

Lesson 5: Flu and Pandemics

To Do:	Date First Completed:
Make a point to start spending 10-20 minutes outside every day without sunscreen. While getting 10-20 minutes of sun exposure every day is the EASIEST thing that you can do to improve your immune system, it's also the easiest thing to forget to do or postpone. One strategy that will help you is to put an "S" on your calendar for the next 30 days. Every day that you get 10-20 minutes of sun exposure, cross through the "S".	
Decide what other (besides sun exposure) preventive measures you're going to take. Buy the necessary supplements and start developing the necessary habits.	
Decide what other (besides sun exposure) treatment options you're going to use if you get the flu. Buy the necessary supplements.	

Lesson 6: Chem/Bio Attacks and "Ghetto" Medicine

To Do:	Date First Completed:
Decide what you're going to do with your pets in the event of an airborne threat.	
Do a test run in your safe room using BLUE painters tape. Monitor everyone's pulse to get used to being able to take vitals quickly.	
If you know that you aren't allergic and have no medical concerns try using superglue on your skin to see how well it works for you and how long it stays on your skin.	
Practice using dental floss to remove a ring from your finger and from someone else's finger.	
Practice some of the duct tape medical treatments.	
Make sure you have duct tape, pads, superglue, safety pins, and wide floss in your 72 hour kits.	

Lesson 7: Teambuilding

To Do:	Date First Completed:
Go over your skills assessment from lesson 4. Update if necessary	
Practice your "half-truth" responses to questions about your level of preparedness.	
Start "casting" or talking with people about survival news that you see or read about.	
Start writing down what you want your group to look like…number of people, proximity, common traits, etc. Share them at www.UrbanSurvivalPlan.com/320/lesson7t	
Make a post on the forum at www.secretsofurbansurvival.com in the "Local Meet Ups" section.	
Check local CERT class schedules at https://www.citizencorps.gov/cert/	
Look into local search and rescue, auxiliary, and mutual-aid programs.	
Sign up for a Clickbank account so you can get paid to promote SurviveInPlace.com > www.surviveinplace.com/clickbank	
When you've got your Clickbank ID, put it into this link and start promoting on Kijiji, Craigslist, and Backpage. www.YOURCLICKBANKID.urbsurv.hop.clickbank.net	

If you take this week seriously and do these exercises and preparations, you will find that you accomplish more this week (and get more value out of the book) than any other week so far.

In the next chapter, we're going to be covering house hardening techniques and the more items you have completed from the first seven chapters, the better.

Chapter 8. Hardening Your House

The concept of "soft" and "hard" targets is straight forward, but if you aren't familiar with it, think of a turtle. If you were going to attack a turtle, would you attack the hard top shell, or the soft underbelly? Obviously, you'd pick the "soft" target over the "hard" target. The same logic applies to selecting an individual house to burglarize in a neighborhood.

As an example, if you're a burglar on Christmas Eve and you see one house with security lighting and a German Shepherd in the front yard and another with a Christmas tree and 20 presents in a picture window with a sign on the door that says, "Grandma, the back door is unlocked. We'll be home from service around 8:00," the soft, easy target is the empty house with the unlocked door and lots of visible presents to steal.

I need to start off right away by saying that you can easily spend hundreds of thousands of dollars hardening your house with upgraded doors, windows, walls, and roofs, but hundred thousand dollar solutions are not going to help most of the people taking this course. Most people who have that kind of money to throw at urban survival would be better off moving to a rural location full-time.

What we're going to do is focus on affordable strategies and techniques that can help protect your family and your belongings now and inexpensive ways to beef things up quickly if you need to in a survival situation. We're also going to get into some heavy topics and cover some more "extreme" strategies at the end of this chapter.

It's important to accept the fact that most houses in the US will always be vulnerable if someone is determined to get in or get you out. Many European homes are built of concrete blocks or large stone. It's a rare house in the US that can withstand a direct impact from a pickup. It's also a rare house that can withstand a few Molotov cocktails or even an ordinary wildfire, as we see almost every summer on the news.

This is why it is important to try to bring as little trouble on your house as possible, both now and in a disaster situation. Maintain good OPSEC, don't escalate conflicts, and don't get involved in fights you don't have to. In an urban survival situation, it's important to blend in and stay as invisible as possible.

We're also going to cover things that you can do that won't be very noticeable to your neighbors. Ideally, we want to make security changes that will deter potential burglars/looters, stop them if they attempt to force their way into your house, but still stay under the radar so you don't get asked questions about "paranoia" or make your house look like a target. We just want to make smart changes that will give us additional security without taking too much time or money.

Let's start by looking at how burglars break into houses right now. Amazingly, 32% enter through an unlocked door or window. According to the FBI's 13,360 reporting agencies, this happened well over 600,000 times in the US in 2007! That was in a GOOD economy with low unemployment.

In another career, I had to get into vacant and abandoned houses to inspect them. Many of the owners were from out of town and waiting for a key to arrive in the mail meant my projects got delayed by several days. As a result, I learned how to get into houses without keys.

I can tell you from personal experience, between sliding doors, unlocked windows, and unlocked doors, it was very rare that I wasn't able to get into houses quickly and without damaging anything.

After unlocked doors/windows, 26% of burglaries are forced by impact, 24% by prying or jimmying, and less than 7% us a pass key or lock pick, including the infamous "lock bumping."

The "where" is as interesting as the "how." 34% of burglars enter through the front door, 23% through a first-floor window, and 22% through the back door. Only 2% entered on the 2nd floor.

And alarms? Well, 41% of alarmed homes where burglaries were attempted had their systems turned off! (You probably found out that your alarm battery dies quickly when the electricity goes out, so unless you have it hooked up to a car battery that gets recharged regularly, don't depend on it in a "lights-out" situation)

Whether you have an alarm or not, it's a good idea to get a lawn sign and some stickers from a local alarm company. In interviews with burglars, most will avoid houses with alarm signs in favor of a "softer" target. Although there aren't statistics on the 41% of alarmed houses that had their alarms turned off, I would put money on the fact that they didn't have an obvious visible sign advertising the fact that they had an alarm in the first place.

So, to harden your house as efficiently as possible, let's go where the numbers tell us to go and take a look at front doors.

Most residential exterior doors in the US could be described as follows:

1. They open inwards. (So they can't be blocked and so you can install a screen door)
2. The hinges are screwed into the door frame.
3. They have a handle and a deadbolt that go into the door frame.

The saying that a chain is only as strong as its weakest link applies here. If a burglar kicks a door and the door is flimsy, then the door itself will break. If it is solid and opens inwards, the force will be transferred to the door hinges and the deadbolt. More specifically, it will be transferred to the screws and whatever the screws are solidly attached to.

Of the two sides of the door, the deadbolt side is normally the weakest since it is doing alone what the two hinges are doing as a team. When you think about how a deadbolt works, it's basically securing your door to your doorframe (instead of to a stud), which is normally only a ¾" thick piece of soft wood. (It's not much different than the boards that get broken in martial arts demonstrations.)

Since doors normally close flush on the inside of the frame, that means that there isn't very much depth to that ¾" piece of wood to support your deadbolt and a good kick is all it takes to get through most doors.

Here are the factors that we can change to make exterior doors more resistant to forcible entry:

Door – If you are replacing your door, go for a solid wood door, fiberglass, or aluminum door, depending on what is recommended for your locale and the direction the door is facing (in relation to the sun.) Check with contractors to see if there are any businesses in your area that stock used doors. It's possible to get great looking, solid, slow growth wood doors that are 50+ years old for less than a flimsy new door.

Windows in doors are great for light and seeing out, but they are also a weak spot on a door and allow a burglar to reach in and unlock/open the door. If you currently have windows in your door, keep reading for how to make them more secure.

Windows by doors – If you have an entryway with windows on one or both sides of your door(s), look into putting security film on them, at a minimum. I'll show you videos on security film on the resource page for this chapter. If you have creative ways to replace the glass with wood that won't detract from the house, please share them with your fellow students about it on the resource page.

Deadbolt – Look for an ANSI level I lock. You can buy levels I, II, and III. Level I is the highest. It should feel nice and solid in your hand…like it could take a few solid strikes with a sledgehammer.

Door Frame – When you buy a pre-hung door, it is pre-hung on a decorative frame, usually wood that gets nailed to the header. If you've ever seen a house being built, you know that the doors get "framed" out way before the doors are actually installed. This framing that creates the door openings is called a header and they are built from 2x4 or 2x6 studs. They are solid and what you want to attach your strike plate and hinges to. If you are installing a new door, consider getting a door that is pre-hung on a metal frame.

Strike Plate – The strike plate is what your deadbolt goes through or into on the door frame. The bigger the strike plate, and the more holes it has for screws to secure it to your studs, the more secure your door will be. Use a flashlight to see how long of a screw you would need to screw the strike plate through the decorative frame and into a stud. In most cases, a 3" screw will go through the frame, through a small dead space, and into a stud.

Hinges – You should use 3" screws with the hinges as well to secure them through the decorative frame and into a stud.

Door Swing – If you are installing a new door, consider installing it so that it swings outwards. It will negate much of the advantage that an intruder has during a forcible entry attempt. Any attempts to kick, ram, or push the door open will be thwarted by the entire door frame in addition to the hinges and deadbolt.

There is a risk in doing this. If you are in blizzard country and get deep snow against your door, you won't be able to open it to dig out. If you are in flood country and water rises quickly, you may not be able to open the door until the water level is the same inside and out. And, if you ever need help from the fire department or EMS and can't unlock the door, they will have to do quite a bit of damage to get in…but since you're trying to make your house difficult for determined people to break into, that's a trade-off you may be willing to live with.

To Do: Remove a strike plate screw and a hinge screw from each door in your house. If they aren't 3" long, then replace them with 3" screws.

Katy Bars. I love the story of the katy bar. King James I of Scotland was being overrun and the queen and one of his maidservants,

Catherine, was with him in his chambers. The intruders were in the castle and fast approaching the king's chambers. Out of devotion to her king, Catherine put her arm where the missing locking bar was supposed to go across the door so that it wouldn't open. She kept her arm in place until the intruders broke her arm, entered the room, and killed the king. (We're going to use materials that are harder than a forearm.)

I'll tell you about a "pretty" solution to this on the resource page, but you can have the materials on hand to make an improvised katy bar in under 5 minutes for $20-$30 per door. Simply buy two "S" or "U" shaped brackets that will hold a 2X4, some 3-3 ½" screws, and a board or metal pipe long enough to span the width of your door, plus the brackets. You don't need to do anything with the brackets now, other than have them on hand. If the need arises, you can pull them out, screw them into your studs on both sides of your door in 5 minutes and drop your katy bar into place.

Power tools with no electricity. Make sure you have a plan for how you will screw and drill if you don't have any power. Many popular tool batteries become almost worthless after a few short years of use. Other than manually drilling/screwing, you can develop a habit of always keeping your batteries charged OR you can look for a 12 or 24 volt drill combo set with a dead battery. Take the dead battery apart and rig it so you can run a wire from the old base to one or two 12V car/motorcycle batteries. It's heavy and bulky but you can charge car/motorcycle batteries with a solar charger and you won't have to worry about it running dead as quick as the batteries that come with your drill.

You can also use an 18 volt drill with three 6 volt motorcycle batteries.

As a note: If you find that the wire running from the battery to the drill gets hot, it means that you need to use a larger wire.

Security Lighting

This is a simple one. Install security lighting with motion detectors covering the main approaches to your house. We have lights with two settings…they turn on at a dim setting when it is dark out and intensify anytime the motion detector detects movement. I have not found any reasonably priced solar security lights that are as bright as 120V ones, so you will have to make a judgment call on which features are most important to you.

Windows

Windows are the next vulnerability that we're going to cover. I am always somewhat amused at houses that have beefed up front doors, but a big picture window 10 feet away that I could easily break and walk through.

Windows are an obvious vulnerability in most houses, but I certainly don't want to get rid of mine and I doubt you do either.

Again, we're looking for solutions that don't make your house stand out, will increase your security now, and increase your security in the event of a disaster.

Plywood is a great field expedient option for protecting your windows in a hurricane, but is obviously a poor choice during normal times. If you use plywood during a disaster situation, it will protect your windows, but it will do so at the cost of reduced visibility. Even so, I would suggest having a few extra panels on hand. Why? Because every other option takes time to research, decide on, and install. Every other option takes significantly more money than plywood as well.

In short, I know that you can go out in the next 7 days and buy a few sheets of plywood, but it may take weeks or months to follow through on any of the other options.

Tempered Glass – While 4X stronger than ordinary windows tempered (safety) windows will still break and allow entry into a house, they just doesn't shatter. This is a good choice for sliding glass windows (required by law) so that people don't get hurt walking through them, but a poor choice for security.

Security Film – Security film is basically a film, like window tinting, that is installed on the inside of the window, so attacks will still break the window, but the film will hold the glass together and keep it from entering your house. This film should be professionally installed, and costs between $5 and $10 per square foot. It will stop bricks, branches, rocks, and bats. If you're willing to pay more, you can even get film that will stop bullets. I've got some great videos on the resource page that show how well security film can protect your windows.

Storm Windows – Storm windows or hurricane windows are laminated at the factory and are made to withstand debris flying into them at 100mph. They are normally tempered and essentially, are made by layering glass and high density plastic sheets. Again, the glass still breaks when it is impacted, but it is held together and there will not be a hole that an intruder can use to enter or reach in. See the video for details.

Lexan – Lexan is a polycarbonate (long chains of carbon) that is bullet resistant or bullet proof, depending on the thickness. It's used for Nalgene hiking bottles and for windows along golf courses. It does scratch more than glass, but is extremely impact resistant. Lexan has a lower R-value than glass and doesn't insulate against sound as well. If you go this route, look for Lexan that is warranted against discoloration, but be aware that Lexan is protected from discoloration by using a UV coating, which is good if you're trying to protect furniture, but bad if you're heating your house with passive solar.

Storm Shutters – These can be accordion shutters, swinging shutters, or rolling shutters. This is a much more expensive option than the previous ones, and more visible, but one that you want to be aware of...particularly the rolling shutters. Popular in Europe for decades, they are controlled from the inside and allow you to have full light, no light, or partial light in a room. They are built out of aluminum, will protect your windows, increase the R-value of your window openings, and are very strong.

Walls

A good friend of mine had a heart-wrenching experience a few years ago that I want to briefly share with you. He was working on a house in a rough part of San Antonio where the houses are small and close together when he heard screams outside. He went out to see smoke pouring out of the house next door. Worse, he saw a mom and her young daughter trapped inside reaching through burglar bars trying to get out. He tried pulling, prying, and everything else he could think of to get them out, but the bars wouldn't budge.

He tried every door and window, got tools from his truck and put his feet against the house and tried to deadlift the burglar bars away from the house. He told me he had tears in his eyes watching the mom and daughter get weaker and weaker.

He kept trying to save them until his hair was singed, his hands were burned, his lungs burned from the smoke, and the mom and child were overcome by smoke and disappeared below the window. This all happened within a few quick minutes, and the fire department arrived as my friend was forced to back away from the house due to the smoke and heat.

They forced their way in with hooligans and pry bars and got the mother and child out, but it ended up being too late.

Here's the tie in to the course…two days later, he went back to work on the house that he'd been working on before the fire. What he saw made him sick. After the fire department broke into the house, somebody re-secured the front door with sheets of plywood.

That didn't stop the neighborhood kids from getting into the house. They simply took a cinder block, beat through the asbestos tile siding, and then beat through the sheetrock and walked into the house between the studs and stole everything that had any value! If my friend would have only known to do this, he feels like he could have saved the mother and child that he watched suffocate to death.

Here's the takeaway…your house is only as secure as it's weakest component, but you can spend a fortune if you're always in search of the "perfectly secured" house. There will always be a "next thing" that you can do to secure your house.

You have to make a decision as to what level of security you're going to be comfortable with and when it makes sense to make upgrades to your house. As an example, it doesn't do any good to have a perfectly secured house if you don't have food and water inside of it.

Decide how secure you want your urban home to be this week. If there are too many things that you want to upgrade, you might want to change your plans or find that it makes more sense to sell your house and buy another home that already has some or all of the upgrades you want in place.

I know a couple who are VERY switched on when it comes to survival preparedness, but who still have vinyl siding, simple doors that swing inwards, and normal windows.

They know the risks involved with this construction and have chosen to accept them. They have an alarm, big dogs that bark, a survival plan, firearms training and martial arts training but just don't have the money to turn their house into a castle.

If money was no object, they'd move to a rural area before upgrading their urban house anyhow. Eventually, they'd love to have a more secure home, but for now they're spending most of their discretionary money on training and travel.

If you're remodeling -- If you don't already have a brick, rock, or concrete house and are planning on re-doing your siding, consider using **Hardiplank or other concrete fiberboard sidings**. In addition to increasing the security of your house, it will probably last longer than you will and it is fireproof.

Landscaping

One landscaping upgrade that you can make today (or in the spring if it's wintertime when you're reading this) is to plant rose bushes or other thorny plants under your windows and in places where you want to stop movement. Make sure to pay special attention to where you plant thorny plants if you have children or children visit you regularly.

If your lot is sloped, you might also consider terracing. If you are into gardening, you already know that this will prevent water and soil runoff, but if the terraces are big enough (2 feet for vehicles, 3 feet for people), it will slow them down and/or force them to follow pre-defined approaches, like your driveway and/or sidewalk.

Up-Armoring Your House

I had considerable debate on whether or not to put this section into the course. I'm very pragmatic and I don't want to sensationalize survival preparations by amazing you with disaster scenarios and preparations that you'll probably never use. If you are in a situation where you need sandbags in a normal American house, other than a flood, you're in big trouble. American homes just aren't made to withstand a determined attack. Heck, most can't even withstand bad storms.

It means that civil order has broken down and someone believes that you have something they want badly enough that they're willing to kill you to get it (or find out if you have it). It probably also means that it's time to relocate as soon as possible.

For the most part, everything that we've discussed in the course has applications in both normal times and in disaster situations. Up-armoring your house is more of a the-end-of-the-world-as-we-know-it strategy and I would suggest reading through it quickly and storing the information but not spending too much time on it. I sincerely hope that it you never have a need to use it.

That being said, here are a few quick, easy, field expedient strategies that you can use to harden your house against small arms fire. To start with, let's take a look at how far various rounds penetrate sandbags/blocks.

Caliber (all FMJ)	Inches of Penetration in Dry Sand	Inches of Penetration in Wet Sand/dirt	Inches of Penetration in Gravel
.22	5		
.223 (M55)	5	6	4
.308	5	13	4
12 gauge slug	5	n/a	n/a
9mm	6	n/a	n/a
.45 ACP	6	n/a	n/a

Notice the difference between wet and dry sand for .308 bullets?

US Army FM 21-75 suggests placing 18" of sand/dirt/gravel between you and your threat. The most obvious way to do this is with sandbags. You can buy sandbags locally through construction supply stores, or you can buy them online. I have a link on the resource page where you can get them for about 50c apiece.

Sandbags are the easiest solution because they are modular, easy to move, and store easily, but here are some other containers you can use:

1. Luggage lined with garbage bags.
2. Rubbermaid (or similar) storage bins.
3. Shirts, jackets, and/or pants, sewn or taped shut.
4. Buckets
5. Using other materials from the course, build an 18" wide trough out of plywood or particle board and line it with plastic sheeting.

Incidentally, it takes 10" of paper (books, phonebooks, etc.) to stop handgun rounds and 20" to stop most rifle rounds.

Again, if it is at all possible, you want to avoid armed encounters, especially in most American homes. If the situation in your area ever deteriorates to where armed violence is inevitable in your immediate area (directed either at you or your neighborhood), it would be smarter (if possible) to relocate to a safer area. You could either make your exit from the city or relocate to a more defensible urban location, such as an industrial/warehouse building, or a stone/masonry house on a multi-acre lot. We'll be covering this in the online resources.

As you're filling your containers with dirt/gravel/sand, one option that you have is to get the material by digging a trench between the street and your house to slow or stop vehicles. It will make it obvious that you are taking defensive precautions, but if you are in a situation where you need sandbags, that is probably a foregone conclusion.

Another barrier that you can use to protect your house and funnel approaches is parking vehicles in your yard. There were families in New Orleans who did this after Katrina to block off their entire neighborhood and it worked quite effectively.

So, in this chapter, we've covered how burglars are actually breaking into houses, simple, inexpensive strategies to make your house a harder target, some advanced strategies you can use to harden your house, and some extreme strategies you can use in disaster situations.

In summary, go with the low cost/high probability fixes first and do the high cost/low probability fixes as time and money allow:

To Do:	Date Completed
Lock your doors and windows.	Every Day
Confirm that your door screws go into your header studs. Replace if necessary.	
Upgrade your locks and strike plates, if necessary.	
Install security lighting	
If you have an alarm, use it.	Every Day
If you have an alarm, make sure you have visible signs advertising the fact.	
If you don't have an alarm, consider getting stickers saying that you do have one.	
Buy wood, brackets and screws to be able to secure your doors and windows in a disaster situation.	
Everything else, as time and money allows.	N/A

Chapter 8 Resource page:
www.urbansurvivalplan.com/365/househardening

Chapter 9. Economics of Survival

In this Chapter we're going to go over some vital SurviveInPlace™ strategies for both short and long term survival. This is meaty stuff, so hold on!

- **Economic time bomb on the horizon**
- Simplest way to build up food that EVERYONE in your family will buy into
- **Food Caches**
- Post-disaster conversations with neighbors
- **Hoarding cash**
- Paying your mortgage after a disaster
- **Paying taxes after a disaster**

I want to encourage you to DO as much as possible in the chapters. As I discussed before, you only retain about 10% of what you read, and when you involve all five senses in the learning process, your retention rate goes up to as high as 90%. The best way to do that is to write out your plan and actually practice survival skills.

When you add in the fact that those numbers will get cut in half again when you're under stress, you need to ask yourself whether you want to remember 5% of the information in this course (reading only) or 45% (reading and doing.)

Timely Information

There are a couple of particularly important items I want to tell you about that are timely as I'm writing this. As I've mentioned in the past, I'm VERY pragmatic. I'm not prone to conspiracy theories, and when people tell me about TEOTWAWKI scenarios, I ask an insane amount of questions to try to boil things down to a plausible level.

First, there is an increasing number of smaller towns that are pre-positioning concrete barricades or have heavy equipment ready to place barricades to block bridges and other choke points to keep "city people" from over-running them in the event of a local, regional, or national disaster. I KNOW of three towns doing this in Idaho, Wisconsin, and Louisiana. Three towns in the entire US is a miniscule number, but these are the locations that I personally know of where the barricades are obviously pre-positioned.

The point is that residents of small towns outside of large population centers know what's going to happen if there's a disaster. They read survival websites too and they know that waves of city people will be heading to the hills (their hills) and they don't want to deal with it. Frankly, they can't deal with it.

Small towns have small town infrastructure and aren't designed to handle a population explosion and know that in order to survive, they're going to have to make the hard decision of turning away city people…even city people who are just passing through trying to get to their retreat locations.

So, again, if you can leave the city before there's a problem, that's your best bet…but if you can't, you'd better be prepared to Survive In Place™ . The main, backup, and tertiary routes to your rural retreat may not be passable due to rural people trying to protect their towns and families.

Second, according to the St. Louis Fed, the money supply has gone up 110% in the last 12 months. There's a link on the resource page to a very disturbing graph that shows the Adjusted Monetary Base from 1918 to the present. In short, the money supply went from $4 billion to $850 billion from 1918 to June of 2008. From June of 2008 to June of 2009, the money supply went from $850 billion to $1.7 TRILLION…more than doubling in 12 months.

During the same time, US gross domestic product has, depending on which report you read, fallen 2-15%.

As a VERY simple explanation, when the supply of money goes up and the production of a country goes down, you have more dollars going after fewer goods and end up with inflation.

I could speculate on what this means from a survival point of view, but the fact is that we don't know what will happen. Never before has a military superpower that held the world standard currency faced a situation like this.

There are a million and one theories including terrorist attacks while our economy is depressed, China dropping the dollar, martial law, splitting the country, and/or a complete economic collapse. Frankly, they are all guesses…some of which are made to scare people into buying stuff and others that could be plausible, but aren't likely to happen.

We don't know what the Fed will do, what Washington will do, what China will do, Al-Qaeda will do, or what any other country will do. I encourage you to stay abreast of what's going on with these theories, but don't get too committed to any TEOTWAWKI scenario.

And, above all, don't give into worry. Be very careful about what thoughts you let linger in your head and only give time and energy to things that you have some control over. You can't effect meetings between national leaders, or the decisions of the Fed, but you can spend more time with your family, exercise more, and spend another hour or two a week on common sense survival preparations from this course.

I was told that the world would end before I got my driver's license, that society would collapse because of Y2K, that EMP attacks were imminent after 9/11, and that the US was definitely going to get hit with dozens of small coordinated terrorist attacks the end of 2008/beginning of 2009. None of them happened. All that I AM confident of is that being prepared to take care of yourself is smart, regardless of the threat and that prices for consumables will go up considerably once inflation starts taking hold.

Food Storage Made VERY Simple

Fortunately, the food storage component of survival will be very helpful any disaster or TEOTWAWKI scenario, including the early stages of high inflation. Keep in mind the obvious fact that if you use your survival food and don't replenish it, it will all go away. The purpose of survival provisions, be it a 72 hour kit or a years worth of food and water, is to help you survive long enough to secure a sustainable source of food and water.

If high inflation hits and causes food prices to skyrocket, you're still going to want to buy some of the food you eat, but you can supplement it with the food that you have stored temporarily until the market settles or you figure out a long term solution to the high prices.

We talked about building up your food storage briefly in an earlier chapter, and if you didn't get serious about it then, I'm going to show you, right now, the easiest way to build up a food reserve, as well as how to frame the conversation with your family to get them on board.

Start by looking at your food inventory from earlier in the course and make a rough guess at how long that would last you in an emergency situation.

Next, if you haven't already, decide how much food you want to end up with. The longer period you are planning for, the more you're going to need to factor in water, vitamins, balanced nutrition, fiber, texture, cooking, flavor, etc. Decide how many people you're going to provide for (family, 1-2 "guests", charity, etc.)

If you have to, you can survive for a long time on beans, rice, and ash cakes (this is the norm in some African and South American countries,) but if you plan ahead, it's simple to have a wide variety of foods that you already eat on hand for survival.

> As a note, my church asked me to put together a plan to feed 500 people 3 meals a day for 30 days (45,000 meals) as inexpensively as possible. You can easily do it for less than $1.79 per person per day using prepackaged, sealed, 5 gallon buckets. This cost includes the cost of propane to heat the food.
>
> If you buy in bulk bags rather than 5 gallon buckets, you can get the prices MUCH lower. If you are interested in this plan for your church or your family, let me know.
>
> On this point, I encourage you to do like I have and talk with the leadership at your church. Briefly and simply lay out the facts about the money supply, GDP, and inflation and ask them if they will consider talking with their staff about building up their family food supplies.

Simple Strategy For Stockpiling Food You Actually Like:

Starting today, when you use a can/bag/box of non-refrigerated food, simply write down that you are going to buy two the next time you go to the store. When you get home, simply put one where you normally would and the other can with your survival provisions.

This prudent exercise is what many call, "Stockpiling." In a disaster, they will call it "Hoarding." Most rural people just call it common sense.

The next time you use a can, replace it with the oldest one from your survival provisions and when you buy your two replacements, put them both in with your survival provisions. Make sure to write the date on the cans as you buy them so you know which to use next.

This simple strategy will work for any non-perishable item, like granola bars, instant rice, seasonings, drink mixes (including powdered milk), all canned fruits, vegetables, soups, tuna, chicken, salmon, non-refrigerated juices and even potatoes if stored properly. It also works quite well for vitamins, disposable contacts, toilet paper & other toiletries.

If you are currently strictly eating fresh produce, then I suggest buying some non-perishable food every time you go shopping. There is no doubt that fresh food is preferable to canned non-perishable food, but I would argue that if inflation causes prices to quadruple, the canned food you have will be healthier than the fresh food you can't afford anymore.

By building up your food storage like this, you will be stocking up on food you eat, that your stomach is used to, and food that you can rotate easily.

A final benefit to doing food storage in this way is that many canned meals are pre-cooked, have water in them, and you can use the can to cook in and eat out of.

One way of looking at this is that if you're consuming approximately 1500 calories a day and you're buying 3000 calories a day, after six months you'll have a 6 month food supply.

These numbers will be off when you factor in meals eaten away from home, but it will give you a rough idea. Obviously, if you can afford to, you can buy 3x or 4x what you use regularly to get your 1, 3, 6, 12, or 24 month supply of food more quickly.

I feel like it is prudent to do this as soon as possible so that IF rampant infiation hits, you've created a safety net for your family. If two or three years go by and we haven't had the inflation that I expect, then you will either have a great supply of survival food or you will have very low shopping expenses when you eat the food.

Getting Your Family to Buy In

This strategy of building up your food supply also happens to be one of the simplest segues that you can use with family members to get them on board with disaster planning. Instead of saying, "I want to buy a $3000 pallet of survival food and put it in our garage," you can say, "Honey, it looks like the price of food and toiletries are going to go up quite a bit over the next 6-12 months. If we can afford it, can you start buying 2-3 extra of everything that won't spoil when you go shopping?"

You don't have to talk about terrorists, TEOTWAWKI, cyber attacks, EMPs, or any other disaster…just talk about buying more of what you already buy. It completely changes the conversation from being a major event to an "Oh, by the way…" conversation.

Another supporting argument for this strategy is to say, "Think of it like a savings account. If we need some more money in a few months, we just eat some of our extra food and use the money that we would have used to BUY food." Note that I said, "extra" food instead of "survival" food. This is on purpose so that the idea is not intimidating.

> **To Do:** Talk with your spouse about inflation and the prudence of starting to buy double of some or all of the non-perishable items that you currently consume.

Cache – It's Not Just A Cool Word In Spy Novels

Picture a scenario with me for a minute. You're 12 days into a post-natural disaster situation. Shelves are bare, even though cell phones started working on day 2, food & water distribution is sketchy at best. A couple of neighbors who you've had over dozens of times come to your door begging for food. They are your friends. You care for each other and want to help them, but you also know that they talk a lot. Here's where it gets interesting.

Option 1: You have all of your food in one location. They've been over for dinner before and they've seen it and commented on it. They know what you've got and know your kids aren't as hungry as their kids. You're going to have to think fast about how to help them without becoming their sole source of food and possibly giving away food that will be necessary for your family in the near future. Again, the concern isn't helping one family with one meal…that's just the right thing to do. The problem is being in a situation where several families come to you repeatedly, feel entitled to your food, and clean you out.

Option 2: You have your food cached in several locations throughout your house. Some visible, some in cardboard boxes marked for something else (crock pot, wine/liquor box, Lincoln logs, etc.) For the last 12 days, you've been using the food from your main location and it's almost gone. Your friends have seen your pantry before and figure you are in the same situation that they're in.

They know how much food you normally have there and this gives you the option to take them to your pantry, show them how little you have left, and offer them something. People in need will generally be more sympathetic to other person in need or even in a slightly better situation than they will be to someone who has "plenty."

By simply splitting up your provisions, you will have made yourself look like you're in as bad of a situation as they are and they will be unlikely to come back demanding more food. You will still have the option to give them food anonymously in the future, or tell them, "I found some Cliff bars in my backpack and knew you guys could use them as much as we can."

These caches don't have to be elaborately hidden caches, especially if your family members are on the fence about survival planning. They should just be hidden from casual view.

> **To Do:** Pick 2-3 locations (closets, etc.) to put caches of food and boxes to put it in.

Will YOUR Water Heater Have Clean Water In A Disaster?

One of the best sources of survival drinking water in urban areas can be the water found in your water heater. Unfortunately, plumbers that I've talked to laugh out loud when I suggest that this is a good source of water in a survival situation.

It's not that water heaters don't have lots of water in them; it's that people make three BIG mistakes when they count on their water heater as a survival tool:

1. Maintenance
2. Bad water from the city
3. Scalding yourself

Maintenance: If you've ever had a water heater go bad and open the drain valve to empty it, you know how much crud (rust & mineral deposits) can build up in a water heater. If you don't drain your water heater regularly (we do it annually,) then you can expect to drink some of that crud if you need to use your water heater in a survival situation, if you can make the valve work at all.

Basic water heater maintenance is very simple, but you can ask a plumber/handyman to help you the first time if you need help.

1. Turn off the water supply.
2. Release the pressure valve until you don't hear high pressure hissing. (You must do this first or the water coming out will be under pressure.)

3. Open the drain valve. I hook a hose up to our drain valve and run it to our floor drain so that I don't get the entire floor wet.
4. Release the pressure relief valve again. This will allow air in the top of the tank and water out the bottom. Drain the tank for one minute or until the water runs clear, whichever is longer. **The water coming out of your water heater is HOT**
5. Close the drain valve. If it won't turn off, and you haven't installed a backup valve, you can simply wait until the tank is empty and screw on another valve until you have a chance to repair the one on the water heater.
6. Turn the water supply back on.
7. Use a magic marker to write today's date on the side of your water heater so you'll know when to do it again.

It's a very smart idea to keep either a screw on Y valve (like you would use to split the flow of a garden hose) or a straight shut off valve on top of your water heater. If you find yourself in a survival situation where you need the water to survive, you don't want to lose the water because the shut-off valve stops working.

In a survival situation, make sure to attach this additional valve to your drain valve BEFORE you start draining it so that you can shut off the flow if the drain valve is bad. While you can also stop the flow of water (if the water supply is shut off) by letting go of the pressure relief valve or raising the end of your hose above the water level in the water heater, the valve is a more reliable solution.

Bad Water From The City: In a flood or earthquake situation, the municipal water supply may not be usable for long. Whether its river water, runoff, or even sewage getting into the water supply, you want to avoid contaminating your "clean" water with "dirty" water. Fortunately, it's simple to do this by either turning off the water supply to your water heater, or better yet, turning off the water supply to your entire house so that you can capture all of the clean water that's in your pipes.

Don't Scald Yourself: After you've drained your water heater a couple of times, this will be elementary, but since scalding is so serious and common with water heaters, I am including a warning about it.

In An Emergency: You're going to use almost the exact same strategy for getting water out of your water heater for survival as you do for annual draining: Screw on your backup valve, turn off the supply, pressure relief, open drain valve, pressure relief until you get enough water, close drain valve. The only difference is that now you won't turn the water supply back on, unless you KNOW that the water coming into your house is drinkable.

As you're draining the water heater tank in a survival situation, turn off the heater if it isn't off already to keep it from overheating as the water level goes down.

> **To Do:** Go through the water heater maintenance steps above. Water from your water heater can burn you. Ask a plumber for help if you are not confident.

Who Will You Let In?: If you face a disaster where there is a breakdown in civil order, you will want to be careful about what you say to people and how you interact with them, but not everyone will be a threat. In fact, there's a good chance that the experience will bring you closer to your neighbors than ever before and possibly even create a small town feeling of closeness in your neighborhood.

One of the things that happened after Katrina is that since neighbors weren't going to work, and since they were going through a common struggle, they talked to each other more, helped each other more, and dropped in on each other more. This will probably be the case in your neighborhood.

That doesn't mean that they need to know about your provisions, and it doesn't mean that they need to come into your house to chat.

In the words of Ronald Reagan, "Trust, but verify." Don't be paranoid that your neighbors are going to take your food and supplies…rather, do your best to make sure that they don't know that you have food and supplies to take.

There's a saying that roughly says, "There are good people, bad people, and people who are on the fence. The good and the bad don't care what you do…they are going to be themselves. But the people who are on the fence can be tempted if you give them the opportunity to get away with being bad."

It is most often used to instruct people to lock their car doors and to keep valuables out of sight, but also applies to dealing with desperate people who may be willing to do things they wouldn't normally do to take care of their hungry families. By simply concealing your supplies, you will eliminate much of the threat.

It all depends on the situation, weather and the layout of your house, but if you have a sitting area in front of or in back of your house, it would be smart to offer to sit down and talk with neighbors that you don't know as well outside of your house rather than bringing them inside. In addition to OPSEC, this gives you control over when you part ways.

If sitting down outside isn't practical, just make sure that you have things that you don't want neighbors to see put away at all times.

Take a second right now and think about what you would do in a disaster situation if there was a knock at the door and it was one of your neighbors. Mentally go up and down your street and decide what your response would be. Who would you invite in immediately? Would you step outside to talk? Talk to them through the door? Talk to them with the door open? Tell them to go away?

What about a stranger? If a crying 20 year old girl is at your doorstep, does she need help or is she the Trojan horse equivalent of a fake breakdown on the side of the road?

What about FEMA personnel, or other "aid" groups wanting to check and make sure your house is safe?

> **To Do**: Spend a few minutes going through these scenarios and decide what you would do in each situation.

The first decision in each case is whether or not to acknowledge their presence once you've identified them. (Is it possible for you to identify visitors without them knowing that you are looking at them? In many houses, it is not.)

Next, decide whether you will talk to them in your house, outside of your house, through an open door, or through a closed door. (Sometimes, if I'm by our back door when someone knocks at the front door, I'll go out the back door, walk around the house, and greet them from behind.)

Finally, decide whether to engage them in a friendly manner or a neutral manner. If at all possible, I would avoid being aggressive. If you are confident, resolute, and polite, you can get your point across without creating an enemy. It's also easy to start off friendly and change the tone to aggressive but VERY difficult to start off aggressive and change the tone to friendly.

In addition to going through this drill mentally, you will most likely be able to go through the process within the next 7 days if you live in an urban area. Even with no-solicitation laws, we get strangers at our door at least a couple times a week and I'm guessing you do too.

Cash: We'll talk about barter and alternative currencies in an upcoming chapter, but for most short term disasters, cash will be the simplest currency to use. Keep as much as you feel comfortable having in small bills ($20 or smaller) in a safe at your house. I would start building this up as quickly as possible.

Many people are unaware, but in the fall of 2008, we were within a few short hours of the entire banking system shutting down. Banks were on a rolling notice and were within hours of being locked and electronic commerce turned off. In light of that, keeping cash on hand is hardly a "crazy" idea. At a minimum, it will save you money if you currently have to pay ATM fees to get access to your money.

Obviously, the fewer people who know that you keep cash in your house, the better.

> **To Do**: Check your classifieds, Craigslist, Kijiji.com, and Backpage.com to see if there are any small fire safes for sale in your area for cheap. Buy one or more if possible.

Foreclosure: In a disaster situation, there are several reasons why you might not be able to pay your mortgage. Your job might not exist anymore, your bank could be shut down temporarily, mail service may not be reliable, or you may not be able to contact your bank by phone or internet.

Keep in mind that banks and most investors don't WANT your house. They want your payments. The foreclosure process is very expensive for banks and is a last resort. After they go through the process, they end up with a house that they have to clean up, pay insurance on, and pay someone to sell.

If for whatever reason, you can't make your mortgage payment next month, the bank is not going to swoop in and take your house immediately. Worst case, if you can't make your mortgage payments in a non-disaster situation, it's going to take the bank 5-12 months to go through the process of contacting you, trying to work something out, sending the case to their legal department, posting a written notice on your door, scheduling the foreclosure auction, and actually selling the house at auction.

In the meantime, they might offer to modify your loan to a lower payment that you can afford just so that they don't have to foreclose.

In a disaster situation, all bets are off. If it's a local or regional disaster, there's a good chance that there will be state and federal assistance, like there was after Katrina. If it's a nationwide disaster, there is no telling what will happen.

I have friends who own several thousand mortgages and car loans. The one constant that has existed through regional disasters, economic booms, and economic busts is that people who communicate with their lender get treated better than borrowers who bury their head in the sand.

So, if you're in a disaster situation and know that you won't have income and won't be able to make your mortgage payment for a month or more, don't panic. Stay in contact with your lender, or at least attempt to make contract and keep a written record of your attempts.

If you get 3 or more months behind and haven't been able to work out a good solution with your mortgage company, you need to start preparing for alternative living arrangements, either by moving in with someone else or finding a less expensive shelter. It's a smart idea to have a backup shelter figured out before you get into a disaster situation.

If you want to negotiate with your bank, one strategy that you can use is to see what houses similar to yours are going for at foreclosure auctions. Calculate what your payment would be if you kept your interest rate the same and owed the bank that price that houses are going for at auction rather than what you currently owe.

As a WAG (wild guess), if your current mortgage is $150,000 and houses are going for $75,000 at auction, ask your bank if they'd modify the mortgage so that you only owe $75,000 and start making payments on that amount. Banks are currently doing this several hundred times a month and will have even more of an incentive after a disaster. It will be better for both you and the bank than going through the foreclosure process.

Keep in mind that while banks don't want to foreclose, it is in their best interest to get as much money from you as often as possible. If there is a widespread disaster, your bank will try to squeeze you for as much as they can, but if the situation is bad enough, they'll be happy to be getting anything.

Taxes: Property taxes after a regional or nationwide disaster are another big unknown, and talking about them is like opening a can of worms. To begin with, the amount that you're taxed has very little to do with the actual value of your home. Here's how they work in most parts of the country:

1. Taxing entities (cities, counties, ISDs, etc.) make their budget.
2. The budget gets sent to the assessor's office.
3. If the assessor can't raise the tax rate (mil rate), they meet the budget by declaring how much properties have gone up in value, regardless of reality.

Is it cynical? Yes. Is it accurate? Befriend someone who works in your local tax assessor's office and find out. Or do a "news" search in Google for "Property Taxes In Florida." I think you'll be shocked at how tame my assessment is. Some Florida residents who saw their homes DROP by 20% in 2008 actually had to pay 15% MORE in taxes in 2009.

I am not a fan of excessive taxes, but I would argue that taxes are necessary to provide for efficient law enforcement, fire departments, EMS, and infrastructure like roads, bridges, water, sewer, electricity, gas, etc.

In the event of a disaster, essential services will still be needed, but if we look at what happened after Katrina, commerce dropped, home prices dropped, and jobs were lost. Tax revenues from sales tax, property tax, and income tax all dropped and the citizens who were left just didn't have the ability to pay local governments enough to cover their budgets.

Cities were able to recover after Katrina due in large part due to federal funds and US Treasury backing of local bonds. This option may or may not be available after future regional disasters and would definitely not be available after a widespread disaster.

What happens in your particular area after a disaster is going to depend on several factors, the main one being what the disaster is. A massive hurricane, earthquake, tsunami, volcano, or local terrorist attack is going to have a much different impact on local governments than a national disaster. Again, with a national disaster, the use of federal funds just isn't going to be an option.

On the local level, the factors include the presence or absence of emergency funds, how many bonds your local government has outstanding and how much your local governments try to provide social services that could otherwise be taken care of by churches, charities, and other private institutions.

It will also depend on your local government's willingness to cut back on spending (non-essential employees) and their ability to come up with creative solutions to compensate essential employees for their time and effort.

What can you do to prepare?

1. Know who you're dealing with as far as your taxing entities are concerned. Living in a no/low income tax state is great when you're making money, but if a disaster eliminates your income, you'd be better off living somewhere with low property taxes.
2. Support politicians in your area who are fiscally conservative.
3. If you have to live in a city for financial/medical reasons and live somewhere with bloated government that isn't making positive changes, consider moving somewhere with smaller state and local government.
4. Look into getting an ag exemption for your home (even in the city.) Many cities offer ag exemptions for residents. Boise, for example, has an exemption for people who sell as little as $1000 a year. This can be $1000 worth of eggs, fruits, vegetables, flowers, hay, etc. and can be done on relatively small lots.

With property taxes, like mortgage payments, you will have more leverage if you are one of a large number of people having a problem paying their taxes and you happen to be willing to communicate and attempt to negotiate your taxes. You will also have more leverage if you have cash that you can use to pay your taxes with, or skills that you can exchange for what you owe in taxes.

Chapter 9 Resource page:

www.urbansurvivalplan.com/382/lesson9

Chapter 10. Alternative Means of Communication

Recent events in the US have shown the telecommunications grid be both amazingly robust and amazingly fragile at the exact same time. I told you about a recent "geek" conference that I was at where the sheer number of people using their iPhones caused 6 AT&T cell towers in downtown San Francisco to overheat and shut down.

I also told you how the major cell carriers brought in portable cell tower/generator trailers and had phones working within a day or two of the levees breaking after Katrina.

Even when cell service was down in New Orleans, and there were Jon boats cruising by in the streets, there were hotels in the French Quarter that had landline phone service, even though you'd think the connections would have been shorted out.

The point is that we just don't know what communication will work after disasters. Cell phones and landlines can be a great primary means of communication, but if they aren't working, it's smart to have one or more backup means of communication. An old saying that has been widely attributed to the US Navy Seals is "One is None. Two Is One." It's not only going to be smart to have backup communication for when your primary communication stops working, it's also going to be necessary to have a simple, efficient way to power your gear.

Email??? One of the simplest "alternative" methods of communication MAY actually be email. Phone lines being jammed doesn't necessarily mean that DSL is down. Even if phone lines and DSL are down, cable, fiber, T1, or satellite internet connections could still be working and allow you to get a message out or communicate with others in your area.

One of the benefits of email, like texting, is that the sending and receiving parties don't have to be engaged simultaneously in order to communicate. So, if you have access now, but someone else doesn't have access to their email for another 2-3 hours, you can still have a conversation.

In addition to actively emailing people, another option is to have an email account with a free email service like Gmail or Yahoo that everyone in your family/group has login information for. When you need to get a message to the other people, simply write the message and save it as a draft. Since it never gets "sent," the chances of getting lost in cyberspace drop considerably.

If you decide to use this strategy, you may also want multiple accounts with Yahoo, Hotmail, and Gmail using the same username and password in case one of the servers is unavailable. You will want to agree in advance about how often everyone will attempt to check these accounts in a disaster situation.

If you can afford to have other people read your conversation, you can also use **Twitter, Facebook, blogs, or even the SurviveInPlace.com forum** to communicate with friends, relatives, your group, or the world. One example of this happened in 2008 when James Buck, a UC Berkley journalism student, took photographs of a demonstration in Egypt and was arrested.

He "tweeted" the word "arrested" on the way to the police station and his friends back in the US immediately contacted UCB, the US Embassy, press organizations, and eventually an Egyptian lawyer who got him out of jail the next day.

Be aware of OPSEC when you post to blogs, forums, and other social networking sites. You have to simultaneously assume that your message will get deleted immediately and that it will never go away.

When I say that you need to assume that your message will be deleted immediately, I'm saying that you shouldn't depend on the intended party ever receiving it. This could be because of posting issues or possibly censorship.

When I say that the message will never go away, I mean just that...Other people will be able to read what you say for days, weeks, or months after you post it, so be careful with your username and be careful with what you post.

VOIP stands for "Voice Over Internet Protocol" and is a technology that lets you talk to people over your internet connection rather than on a regular phone. Some commercial names are Vonage, Skype, and Magic Jack and they cost as little as $30 per year for service. If you have good enough internet access in a disaster situation, you will probably have VOIP access and be able to talk with other VOIP users and/or call telephone numbers outside of the affected area.

The major benefit is that your VOIP phone will be able to send and receive calls anywhere that you have a high speed internet connection...not just at a fixed location.

That's great…how do I connect to the internet after a disaster?

Strange things happen in disaster situations. Guests in the hotel I mentioned that had simultaneous flooding and working landlines ASSUMED that the lines were down and didn't even try making calls for one to two days.

Even if the internet connection at your house is down, it's very likely that with all of the different methods of connecting to the internet that you will be able to find one that works after most disasters. Here's how.

War Driving is the practice of finding unsecured wireless (WIFI) networks that you can "steal" bandwidth from to surf the net, talk via VOIP, download email, post pictures, comments, tweets, etc.

There are handheld devices made specifically for "sniffing out" unsecured WIFI connections, but you can do the same thing by driving around with a laptop in your car.

If you've got a laptop and have never done it, take a drive with your laptop turned on and your "View Available Wireless Networks" tab opened. In most residential areas, you'll have the option to log onto a dozen or more networks per mile and in high rise areas, it may even be more. In fact, there's a good chance that one of your neighbors has an unsecured connection that you could access right now, although I don't suggest that you do that since it is technically illegal.

Depending on the disaster that you're facing, you could find that your cell phone doesn't work, but that you can connect to someone's DSL, Cable, or Satellite WIFI network with your laptop or WIFI enabled phone, allowing you to access email, tweet, or post to a blog.

In many cities around the world, war drivers have been marking unsecured WIFI locations for several years. I'll link to some guides on the resource page for this chapter at www.urbansurvivalplan.com/441/lesson10. The idea behind these markings is that if you find an unsecured network, you mark the sidewalk, a wall, or lamp post near where you get the signal so that other "war drivers" can access the network in the future.

The ability to be able to quickly send/post pictures that you take on your phone in a disaster situation could be very valuable and I encourage you to practice this if you have a WIFI enabled phone with a camera. Many coffee houses have free, unsecured WIFI and they will allow you to try out this skill without stealing anyone's bandwidth.

> **To Do**: If you have a WIFI enabled phone with a camera, practice taking a picture, logging onto an unsecured public WIFI network, and sending the picture by email. Ask someone you know who has advanced computer/phone skills to show you how if you have any trouble.
>
> If it is challenging for you to do, write down the instructions and put them in your SurviveInPlace™ plan.

Another reason to practice this is so that you can see how vulnerable unsecured networks are. If you have an unsecured network at your home or office, someone could easily do illegal acts (view kiddy porn, use a stolen credit card, harass a public figure, post hate speech, etc.) using your network. If the police investigated the incident, they would see that the incident originated at your IP address and you would be the primary suspect. You would most likely be found innocent eventually, but would have to prove your innocence in the meantime.

As I talk about in my book, "**47 Proven Identity Management and Identity Theft Prevention Techniques,**" the default admin login information is easily available online and anyone who wanted to could also lock you out of your own network in a matter of a couple of minutes if your network is unsecured.

> **To Do:** If you have a wireless network, make sure that it is secured and that it does not use the default admin login. I write the updated admin login on a post-it note on the routers that I have so I won't forget it in the future.

Two Way Radio Communications

FRS (Family Radio Service) & GMRS (General Mobile Radio Service) Radios are a cheap, easy backup for cell phones and they're what I have used on road trips, guiding backpacking trips, and on close protection details. Some models claim that they have a 10 or 20 mile range, and while this is technically possible, I have not found it to be practical. In order to get this range, you would need to basically be in a flat dessert with no physical obstructions between radios, not have any radio interference and be at different elevations (or both be on hilltops with valleys in between) so that the curvature of the Earth doesn't block the signal.

Yes...curvature of the earth limits FRS and GMRS (and eXRS) radios to a maximum effective range of 6 miles, despite what advertising in the stores claim.

These radios do work well over short distances in urban areas...even through some buildings. In short, you need to test your radios before you use them so that you know how they will perform in urban and wilderness areas.

When you're buying walkie talkies, I suggest buying ones that take both rechargeable and ordinary AA batteries, like the Motorola Talkabouts so that you have multiple options for power.

If you're concerned about range with a GMRS radio, I have links to external antennas for your home and car that will get the antenna higher and extend your range considerably.

CB (Citizens Band) Radios are an option, but I would encourage you to use FRS or GMRS radios instead due to how much larger CB radios are and how simple FRS & GMRS radios are to power.

Short Wave, MURS, VHF, Wired Radios and Repeater Radios are a HUGE topic that are far beyond the scope of this book. Suffice it to say that they can be powered from car batteries, made portable, and short wave can even communicate half way around the world. All of these are GREAT survival communication tools, but they all take much more time and research to make a decision on than FRS and GMRS.

This is similar to the saying that "The 9mm that you're carrying is better than the .45 at home in your safe." In many cases, one of these other radios would be "better" than FRS/GMRS radios in terms of range, fewer people on the frequencies, etc., but they aren't systems that you can go to your local store, buy, buy AA batteries for, and start using them immediately for under $100 TOMORROW.

At a minimum, I'd suggest getting a set of FRS/GMRS radios for backup two way communication and a crank/solar powered FM/AM/short wave receiver like the Kaito Voyager. It will allow you to listen to shortwave radio broadcasts and news that could be much better than your local news in a disaster situation.

OPSEC. Anyone with the same type of radio can hear your transmissions on these radios, so watch what you say and decide in advance what names you want to use and any code words, if applicable.

One strategy for names that I came up with and have used successfully is to assign everybody on your team/in your family a letter. It could be A,B,C, or it could be "D" for David, "P" for Peter. In either case, the first letter is all that matters.

Let's say that I am assigned the letter "A." It means that I will answer to ANY name that starts with the letter "A." It could be Adam, Aaron, Andy, Andrew, Amy, Angie, etc. I can also identify myself using any name that starts with the letter "A." One advantage of this is that it doesn't sound like code. I'm not saying, "This is Alpha," or "This is Maverick." I just say, "This is Adam."

With non-secure communication, you want your conversations to sound as un-interesting as possible so that other people will simply change to another channel rather than listen in.

Some codes that you may want to decide on ahead of time:

Danger. Come and help.

Danger. Flee. ("Look at that balloon going up outside of the bank")

Change channels to the next pre-determined channel.

Come to my location discretely.

Something suspicious is at my 3:00 (straight right), 15 meters out. (I've used a system like this: "Would you rather meet in 15 minutes or at 3:00?")

Try it out at a mall, a market, a theme park, or a fair. Again, in many cases it is preferable not to sound serious or tactical and to simply sound like friends talking.

Graffiti?? As I mentioned with war driving, there is a graffiti communication system in use to communicate information about wireless networks. It is actually based on a decades old system called "hobo chalk codes." It could be very useful in a survival situation.

From denoting which dumpsters have good food and who's an easy mark to where hobos get beat up, where homeless stings are happening, to which stores/homes have armed men, the hobo chalk codes can help urban survivalists recognize opportunities and avoid danger.

I've got both war driving and hobo chalk code guides on the resource page for this chapter. Make sure to take a look at them and even consider printing them out and putting them in your SurviveInPlace™ Plan.

> **To Do**: As you're driving about, see if you can spot any war driving codes or hobo codes. They are much more common in the northeast and on the west coast than other places in the country, but you may see some. As you see them, snap a picture and send it to me.

Satellite Telephones is one final two way emergency communication method that I'll discuss. They have such strong good points and such strong bad points, that I can neither completely recommend or completely discourage you from using one.

On the plus side, they won't be subject to local infrastructure breakdown or jammed phone lines due to a local disaster and you can use them in wilderness areas and out at sea where you can't use cell phones.

On the downside, they're expensive, they don't work in some urban areas where cell phones do (underground parking garages,) and if you're concerned about conspiracies, satellites can be shut down as easy as a cell network and conversations easily monitored.

> **To Do**: Decide what forms of backup disaster communication your family will use. If you use radios, decide on channels, names, and codes, as well as times that you will turn on your devices if you are low on battery power.
>
> If you are going to post to a website or use a common email account, make sure everyone knows how to post to it.
>
> Write this information down and record it in your SurviveInPlace™ Plan.

How Do I Power This Stuff???

Your basic options are batteries, cranks, solar, or generators. I'm going to cover generators in the Katrina lesson, so today we're going to discuss cranks, solar, and batteries. The big advantage of cranks and solar is that, unless they break, they will continue to work for years. They won't die like batteries or run out of gas like generators, so I consider them a vital component in most people's survival planning. Unfortunately, they're not real simple.

You may be saying to yourself, "The Native American Indians didn't have batteries." Or "Les Stroud doesn't use batteries on Survivorman." You're right. People who are highly skilled in survival do not NEED batteries. They don't need headlamps, flashlights, radios, or communication.

If you're like me and want as many advantages as possible in a survival situation. You're going to want to be able to communicate with others, have flashlights at night, and do other activities that require batteries.

I've got to get a little bit technical on you for a minute for all of this to make sense. You see, when you go shopping for solar and crank powered survival tools, you'll see a LOT of grandiose claims.

****begin major geek speak****

To help you understand what all the terms mean, I need to explain what watts, amps, and volts are. To use a common analogy, think of a garden hose spraying water. The amp rating is the volume of water flowing through the hose. The voltage is the pressure, and the wattage is the total amount of water that comes out of the hose per unit of time. They relate to each other through the equation Watts=amps X volts.

AA batteries are rated at either 1.2 or 1.5 volts and most alkaline AA batteries have around 2000 mA and NiMH rechargeable AAs have between 2000-2500 mA (milli amps) when fully charged. (AAA batteries have roughly ½ as much)

Milli amps are simply 1/1000ths of an amp. 1000mA=1Amp

Using the above information, a 1.2 volt fully charged 2500 mA AA battery will hold 1.2 Volts X 2.5 Amps = 3 Watts of power and would power something that requires 3 watts for about an hour.

****end major geek speak. Begin minor geek speak****

One popular compact solar battery charger, for example, advertises that it puts out 80-160 mA of power. It claims that it can charge 4 AA batteries in 3-6 hours. When you dig deeper, their claim falls apart.

The batteries that the company used for their test were 500 mA batteries, rather than the 2000-2500 mA batteries that you're used to using.

Put another way, it would REALLY take 31 HOURS of direct sunlight to charge "normal" rechargeable AA batteries using this charger, rather than the 3-6 hours that they advertise. (2500 mA / 80 mA/hr = 31 hours)

Crank chargers are somewhat better when it comes to compact chargers. The Kaito Voyager, as an example, has both a solar panel (4.5V 40 mA/hr max) AND a hand crank (5-6V and 500-600 mA/hr at a 2 crank per second pace.) This is better, but it still means that you would have to crank for just over 4 hours to fully charge 4 2500 mA rechargeable batteries.

(This is a great little radio…I have one and I fully recommend it as an emergency radio, but do not recommend it for charging batteries/appliances except for limited use.)

There are four big takeaways here:

1. Test out your survival equipment before your life depends on it. Don't just assume that since it says it can "charge anything with a plug" that it can do it in a timely manner. If you aren't WILLING to crank the equivalent of a pencil sharpener for hours at a time right now when things are good, don't plan on being ABLE to do it in a disaster situation. Find a more realistic solution.

   Also, if you don't have DAYS for the sun to charge your batteries, make sure you have a big enough solar array to charge your batteries as quickly as you use them.

2. You've got to know a little math when you're buying solar and crank chargers. Here are the biggies: Good AA NiMH batteries hold 2000-2500 mA. Divide that number by the number of mA of the charger you're looking at and you'll know how long it will take to charge the batteries. If you only drain the batteries part way before recharging, adjust your numbers accordingly. In any case, you need to make sure that you're USING fewer mA per day than you're GENERATING to have sustained power.

   Watts=volts x amps. If you're charging 4 1.5 volt batteries, then you need 6 volts of input. If the charger that you are looking at is rated in watts instead of mA and is rated at 2 watts, it means that you're putting 2 watts/6 volts=.33 amps or 330 mA per hour into your batteries. If they're completely drained, it means that it will take 2500 mA / 330 mA / hour = 7.5 hours to charge them.

Most output ratings on solar chargers are "ideal" and you will only get that output by continually pointing the panel towards the sun throughout the day. If you don't continually reposition your solar panel, you should only expect a maximum of 3-4 hours of fully effective sunlight per day.

3. For short term emergencies, you could consider just going out and buying a LOT of regular batteries. As an aid in figuring out how many batteries you could need, Motorolla talkabouts go through 3 AA batteries every 30 hours. LED lamps vary widely from a couple of hours to 1200 hours (50 days.) I would suggest buying a couple of 30 packs of each size of battery that you regularly use, if not more, and cycle through them, just like you do with food.

4. In interviewing solar/generator experts, one consistent bit of feedback I received was that 90% of people who start out looking at solar off-grid solutions or solar backup systems end up going to Costco and buying a generator.

****end of geek speak****

Hopefully I haven't scared you out of using solar or crank power. They are great options in a survival situation, but you do need to know the limitations and size them correctly for the application. With that in mind, I want to tell you about some options that will give you enough usable power to help you in a short to long term survival situation.

This technology changes very rapidly, so I won't be mentioning models, or places to buy the following items here, but I will have examples on the resource page for this chapter so that you'll always have access to the newest information. I'll be updating the information as I continue to do research and get feedback from students, so if you find a better vendor, better pricing, or a better package, make sure to let me know!

If you need to recharge high capacity rechargeable batteries quickly, you're going to want a bigger system than what you can get at camping stores.

There are several systems available that cost around $200, put out 12 watts, and can charge 4 high capacity AA batteries in 4-5 hours.

Put another way, they will charge batteries at 600 mA per hour.

Since many of them have a female cigarette lighter plug, you can charge anything that has a car charger.

The next option is a big step up in both price and features. For around $600, you can get a 40 watt solar array, a 55 amp-hour deep cycle battery, 300 watt inverter, a charge controller, voltage meter, and a 12 volt DC outlet. Again, to figure out exactly what that means in terms of AA batteries, we take watts=volts X amps.

If you're recharging 4 AA batteries (6V), then 40 watts will charge them at a rate of 6.6 Amps per hour and your batteries will be fully charged in about an hour. There is enough power to charge them faster than that, but most batteries can't take being charged any faster than that.

The big advantage of systems like this is that the lead acid battery holds the equivalent of 220 AA batteries of energy when it's fully charged. You can use it to power anything that plugs into a cigarette adapter. With the included 120 V inverter, you can even plug in small household appliances. (More on lead acid batteries in a minute.)

If you're interested in a small, foldable version of this setup, Brunton has several large solar arrays to choose from, but they cost 2-3 TIMES as much as the ones I'm showing you for the same amount of power and storage.

As you know, the sun doesn't always shine. Whether it's because of storms, smoke, clouds, or your location in relation to the equator, solar power doesn't work in all situations.

Fortunately, there is a simple solution to this problem that will work for people who are in good enough physical shape to pedal. (If you're not, I suggest making that a goal. If medical conditions make that impossible, try to find a "young buck" or two to add to your team.)

One energy generation solution that has been around for several decades is simply attaching pedals to a generator. You can build your own if you are mechanically inclined and I have links to plans on the resource page for this chapter.

You can also buy pre-assembled systems that will work with a bicycle. As a bonus, they can be used on their own or in combination with the solar/battery system above. A pre-built pedal system that a moderately fit person can use will put out 150 watts for around $300. This is almost four times more than what the solar array will put out. Again, I have links to a couple of options on the resource page.

Batteries are not created equally. One of the solar experts I talked with for this chapter has been helping people with off-grid homes and power backup systems for 40 years. I originally approached him about buying a couple of car batteries and a 2'x4' solar panel. I had the brilliant idea that I could have an "even" battery and an "odd" battery. On even days I'd charge the even battery and use the odd one. On odd days, I'd charge the odd battery and use the even one.

After I explained what I wanted to do for the 2nd time, he laughed (nicely) and proceeded to tell me that I was a complete idiot. (again, very nicely.)

It turns out that car batteries are a very poor choice for off-grid power, but it's a common mistake to try to use them for that purpose. Car batteries are designed to work for a very short period of time (cranking the motor) at high output and then immediately charged back to a full charge. For this reason, they are called "shallow cycle" batteries.

If shallow cycle batteries are drained more than 20-30% their life spans will be shortened considerably and if they are left in a drained state for more than a day or two, they can be completely destroyed.

Deep cycle batteries, on the other hand, are made to be discharged as much as 80%. Rolls batteries, for example, can be discharged 80% and have a lifespan of 7-15 years. A few non-survival example of deep cell batteries are golf cart batteries, RV "house" batteries, and marine batteries, although not all of them are truly "deep".

What's the difference? Here's a simplified explanation…a lead acid battery is made up of lead plates dipped in acid. The more surface area of the lead that touches the acid, the more amps the battery has.

The lead plates in shallow cycle batteries are like sponges and have a lot of surface area, but corrode quickly if they aren't kept charged. Lead plates in deep cycle batteries, on the other hand, are big, solid blocks of lead. They don't have as much surface area, but they can handle being drained much deeper.

The other big difference is price. High end (Rolls) 12 volt deep cell batteries run up to $1000 apiece, but you can buy MK deep cell batteries for as little as $100.

If you decide to go this route, your battery and/or charger will come with maintenance instructions. Make sure to follow them so that your battery lasts as long as possible, doesn't explode, and doesn't leak acid on you or your valuables.

As a basic cautionary note with any car, marine, or deep cycle battery; storing batteries on concrete floors will cause them to die very fast.

> **To Do:** Decide what power solutions you are going to use in a disaster situation. Add the items to your SurviveInPlace™ prioritized supply list. At a minimum, start buying extra batteries in the sizes that you use when you find them on sale in bulk.

That's it for this chapter. It's a fair guess to say that if you can go over this information a couple of times this week and retain 20-30% of the information, you will know more about emergency communication and batteries than 95% of the population. I know it was a little more technical than usual, but I found the information necessary for my own preparations and I hope you agree that it was valuable. Before I knew the numbers on solar and crank chargers, I thought that I had the ability to easily recharge all of my electronics with my pocket solar charger and a hand crank radio. As you now know, the facts are a little different. Hang tight...we've still got some important topics to cover, including:

Lessons from Katrina...from security contractors and people who teamed up and rode out the storm.

Sleep, and psychology...everyone has to face these things every day. We'll walk through how to do it in a disaster situation.

Venturing out...safe travel and commerce.

Chapter 10 Resource page:

www.urbansurvivalplan.com/441/lesson10

Chapter 11. Lessons Learned From Katrina

Natural Disasters vs. Man-Made Disasters

Katrina illustrated one of the unfortunate new realities that we have in our entitlement-minded America: The response of criminals and entitlement-minded people who have nothing to lose to a disaster may be worse than the actual disaster.

There is no doubt that Katrina had a devastating effect on Louisiana and Mississippi, but many of the biggest lessons that we can learn from the event had to do with what happened afterwards. Most of the horror stories of Katrina were man-made. On one hand, that fact sickens me, but on the other hand, it means that the lessons can be applied to all disaster situations… and that the people who went through the experiences didn't do so in vain.

Timing is Everything

The first lesson that we can learn from Katrina is the importance of timing. Regardless of the disaster, if you decide to bug out, you must make sure that it is still a practical option before you actually leave the driveway.

People who left New Orleans early had no problems at all. The roads were clear, gas stations had gas, food, and drinks, and hotel rooms were available a few hours away.

As an example, in the days before Katrina made landfall, the travel time from New Orleans East across the Twin Span Bridge was only 15 minutes.

As more "last minute" evacuees started leaving town, the travel time increased to 17 HOURS. The problems weren't "linear," but rather compounded as time went on as a result of cars running out of gas, accidents, and a lack of available fuel. From talking with people who were there, it was like someone slammed a door shut.

There were several factors that compounded the problem, but one of them was that many evacuees started out with near-empty gas tanks and were prepared for a 15-20 minute drive across the bridge. They were not prepared for 17 hours with their engines idling. As their cars ran out of gas, they blocked the cars that DID have gas.

One of the biggest factors came into play between T-24 hours and T-12 hours. It was during this time that almost every gas station within 100 miles of New Orleans ran out of gas (and stopped getting resupplied).

One man reportedly paid $1,000 for a tank of gas during this time. In addition to running out of gas, convenience store supplies were wiped out (including toilet paper for the bathrooms).

If you are ever in a mass evacuation like this, it will be important to keep your food/water close to you and hidden. Also, eat/drink discretely so that you don't get wiped out by people around you who are in need and see that you have food and water. If your vehicle permits, it would be smart to have multiple small bags/containers of food/water so that, if necessary or advantageous, you can show people a single small bag/cooler and appear to have very little.

If you have a problem with the mentality of not sharing your survival provisions with other people who are in need, that is fine. Just make sure to game it out in your mind in advance. You need to decide whether it would be easier for you to watch your loved ones or strangers complain that they are hungry/thirsty. Don't wait until you are in the middle of a disaster situation to address these questions. Predetermining your decisions for these kinds of moral dilemmas when you are well fed and rested will allow you act decisively and efficiently when you're operating under stress.

As an example, imagine being stuck on the Twin Span Bridge for 17 hours. People will not live or die of dehydration based on whether or not you share your water during a 17-hour situation. They may "suffer", but they will not die if they keep their heads screwed on.

On the other hand, YOUR strength, YOUR ability to see, YOUR brain's ability to communicate with YOUR muscles, and YOUR brain performance will start dropping noticeably when you are dehydrated by as little as 2 quarts (or liters) if you weigh 100 pounds or 4 quarts (or liters) if you weigh 200 pounds. Keep in mind that you will lose this amount of fluid to sweat and bodily functions in as fast as 1-3 hours if you are hot and under stress.

The same rule applies to food. People around you may "suffer" some if you don't share your energy bars with them, but they've got weeks before they are in danger of dying of starvation. On the other hand, if you run out of food because you shared too much of it, you are going to start experiencing an increase in irritability and a loss of mental and physical function due to low blood sugar.

The loss of "high speed" functioning due to either dehydration or low blood sugar is not something that you want to have happen in a survival situation...especially if you're surrounded by hungry, thirsty, irritable people.

Just remember that the people around you made a conscious or unconscious decision to NOT be prepared, just like you made a proactive decision to be prepared. Unless you are a few days into a disaster, giving away your survival provisions won't make the difference between life and death for someone else, but it will hinder your ability to operate at 100% if you run out of food or water.

This doesn't mean that you shouldn't share anything with anyone. If you have your provisions split up, you can easily show someone a small cooler with a few bottles of water, fruit, sandwiches, bars, etc., and tell them, "I don't have much, and I don't know how long we'll be here, but you can have some." You can still be a "good guy"—just do it smartly.

Don't Believe Everything That You See/Hear On TV

Many of the breakdowns of post-Katrina were related to communications, and one of the biggest problems was misinformation spread by the media.

TV, radio, blog, and newspaper reporters were all too happy to publish any horrible story that they were told. HORRIBLE things happened after Katrina, but nothing like what was reported in the news. There were reports of horrendous crimes at the Superdome, including babies with slit throats, but they were never found.

There were also reports of "thousands" of bodies floating in the streets, "snipers" shooting at helicopters, and cannibalism.

There WERE bodies floating in the streets, and I have a link on the resource page to pictures, but there weren't "thousands." There were stupid thugs shooting at helicopters because they were stupid thugs, but they weren't "snipers." And there were no substantiated reports of cannibalism.

Unfortunately, there was a real consequence to this irresponsible reporting. Steve Sailor from iSteve.com said it best with this observation,

> "Sure, rumors outrun the reality, but think about what it would be like to be a cop or fireman who is supposed to go out in a boat and rescue people. You're putting your life vest on because there's a chance that some desperate survivor in the water might pull you in. But then your wife rushes in and says there are reports of ~~snipers~~ (thugs) shooting at rescuers, and she insists you put on your bullet-proof vest instead. But that's heavy and would drag you right down to the bottom. So, you say, screw it, I'm calling in sick." – Steve Sailor, iSteve.com
>
> *(changing "snipers" to "thugs" was done by me out of respect for true snipers.)*

If you were a police officer and you knew that this kind of activity was going on, would you stay home and protect your family and your neighborhood, or would you go out and try to protect people who might thank you by shooting you?

Compound this with the fact that 80% of New Orleans Police Officers reportedly lost their houses to the storm and subsequent flooding and 2/3 of them basically quit the department after the levees broke, and you've got a serious staffing problem.

These news reports had consequences. They emboldened criminals, they broke the already fragile will of much of the New Orleans Police Department, and they made ordinary citizens more scared than they needed to be.

The takeaway here is that you need to have a good mental filter in place when you listen to the news after a disaster. Don't believe everything that you hear, and don't let news that you don't have any control over affect you.

Remember that it is the job of TV and radio stations to keep you tuned in for as long as possible, by whatever means necessary. It's not their job to help you survive or to give you accurate, actionable information that you can use to survive. It's a bonus when they do this, but their primary purpose is to keep eyes and ears tuned to their station and watching/listening to their ads.

It's important to realize that all of this mis-reporting was most likely the result of simple sensational journalism and a thirst for ratings rather than a coordinated, malicious effort of putting out mis-information. We'll discuss this more when we cover the mental aspect of survival, but if you don't have control of how your mind deals with news, it will react to sensational journalism as if it is true, regardless of whether it is or not.

Adapt and Overcome

Seventeen television stations and 79 radio stations were forced off the air by Hurricane Katrina. In an example of adapting and overcoming, FEMA still managed to get pertinent information out to residents telling them about dangers to be aware of, locations of shelters & aid stations, and how to deal with waste/water issues. How'd they do it? By printing up flyers and delivering them door to door.

Unfortunately, in what looks like a rare success for FEMA, they screwed this up as well. You see, firemen from across the country volunteered to go to New Orleans and Mississippi to help out in the wake of Katrina. FEMA decided that they all needed to get "trained up" in Atlanta before being "deployed" to their assignments.

Their training in Atlanta included 2 days of training on workplace sexual harassment issues and when they did get deployed, these firemen were tasked with handing out flyers rather than rescuing people or providing medical care.

Don't Expect Help If Communications Are Down

While some of the communication problems involving the media made the aftermath of Katrina worse, there were many other communications issues that caused problems as well and served to emphasize the point that you need to be able to take care of yourself after a disaster.

Starting at the local level, 52 emergency (911) communication centers were disrupted. Some had to be evacuated due to flooding and some simply did not have adequate backup power. This was the result of not having generators, having too small/few generators, not having enough fuel, or having the generators in locations susceptible to flooding.

Even in the areas that DID have operable 911 centers, many radio repeaters were inoperable, crippling police, fire, and EMS communication.

> Quick clarification: Radio repeaters repeat radio transmissions. Police, fire, and EMS use repeaters so that a responder can use a relatively weak radio to communicate with a dispatch center that may be several miles away. When they start transmitting, their signal is picked up by the repeater and rebroadcast from a higher elevation and with more power.
>
> Cross band repeaters do the same thing, but they allow agencies on different frequencies to communicate (fire/EMS/police/feds/different jurisdictions).

With 911 centers, repeaters, cross band repeaters inoperable, and many vehicle mounted radios flooded, interagency communication stopped as well.

In the days when most regular emergency communications were down, four of the most reliable forms of communication used were courier, shortwave radio, satellite telephones, and plain old family radios. Many agencies did not have enough shortwave radios, satellite telephones, family radios, or batteries on hand.

As State, National Guard, FEMA, Coast Guard, private security, and outside first responders started to arrive, and shelters started to be set up, the communications problems intensified. Not only were the different agencies operating on different frequencies, they used different jargon when communicating. Among other problems, local people used landmarks when giving directions that had no meaning to outsiders.

I'm telling you about these problems for a couple of reasons:

1. After a disaster, assume that you're on your own, even if the city is full of police, fire, and EMS.
2. If someone (police, fire, EMS) tells you they're going to contact someone to help you, don't believe it until you see it. The intent may be there, but in reality they may not be able to guarantee help with so many constantly changing variables.
3. If you have the tools and skills and choose to volunteer, you could be of tremendous value as a radio specialist who can bridge the communication gap between agencies.

In addition, outside first responders and private security will likely need local communication help, and relief shelters will need assistance reaching EMS & police if they don't have the appropriate equipment. A good parallel to this is the relationship between private security forces and NGOs (Non-Governmental Organizations) in war zones. The NGOs typically want to come in and "help," but they need/want a local to help them with local knowledge, communications, and security.

Good Luck Getting Medical/Trauma Help

For several days after Hurricane Katrina hit New Orleans, only one hospital, Ochsner Medical Center, was able to take new patients. Other hospitals were used as evacuation points, but weren't able to accept new patients due to staffing, water, food, and electricity issues.

Built in 1951, the founders of Ochsner Hospital knew that it was only a matter of time before the levees would fail, water would be compromised, and the city would NEED a working hospital. As a result, Ochsner drilled its own redundant wells, installed redundant generators that were higher than the levee, and built up food stores. Most importantly, the entire structure was built up high enough so that flooding from the levees would not be an issue.

(If you work with a company or organization that needs to keep working effectively after disasters, Ochsner is a great case study. They supplied food, & lodging for employees and some families. Not to mention that they figured out how to handle regular payroll for those working in advance as well. They ramped up staffing for multiple shifts before the storm so that they could stay operational through the storm, after the levee broke, and until order was restored. It also is a great study into the cost of doing the right thing after a disaster. Ochsner provided over $80 million in un-reimbursed medical care after hurricane Katrina, even though other parish hospitals that were not operational got "reimbursed" by FEMA.

Most other hospitals in New Orleans and Mississippi were not so well prepared.

One of the big problems that they faced was recordkeeping. Some hospitals in Mississippi lost ALL electronic and paper record keeping capacity. What this meant is that 49% of the patients entering the EMS system were not able to be tracked by family members and there was no record of their prior care/treatment/diagnosis as they moved through the EMS system. Many were moved by the busload to other states without tracking or without notifying family.

One close protection specialist from New Orleans told me about one of his friends who got hurt (in a coma). His girlfriend of 7 years took him to a parish hospital for help, but wasn't able to get any updates on his condition after he was admitted. Even though they owned a house together, the hospital refused tell her the condition of her (for all intents and purposes) husband, or even where he was. It turned out, they transported him to another state almost immediately without telling her. It took her over a week to find someone who was willing to break the law and tell her how he was, where he was and how to see him again.

Once patients got to "evacuation hospitals," fuel shortages caused waits of several hours to a few days for transportation to a hospital out of the damaged area that had the space and manpower to actually admit them.

In all cases, whether operating as a hospital or as an evacuation point, hospitals had severe security problems dealing with the "walking worried" who were looking for friends and loved ones. Lack of accurate records meant that the only way to know whether or not the person you were looking for was at a particular hospital was to go looking for them.

In most cases, these walking worried were honestly looking for family and friends. In other cases, they were thugs who used the opportunity to steal drugs, supplies and patient belongings. In all cases, the sheer number of people passing through caused patients to have a lower level of care and increased their chances of catching secondary infections.

Hospitals had as many as 20 times more patients as normal and with their normal supply channels cut; they quickly ran out of drugs and supplies. With the help of local law enforcement, some hospital employees improvised and overcame the problem by scavenging local pharmacies until shipments from outside sources started arriving.

Why am I telling you about breakdowns in the medical system? Because in addition to serving as another warning not to depend on others after an emergency, I want to stress the importance of avoiding activities that could put you in the hospital, such as physical conflict, driving in areas without traffic signals, running out of needed medication, and physically exerting yourself if you normally live a sedentary life.

By practicing conflict de-escalation, slowing down/stopping at intersections, pre-planning your medication, and staying physically fit, you can avoid many of the most common reasons that caused people to need EMS care after Katrina.

Stay Away From Shelters

I am going to start out by saying that I'm not going to be critical of the people who ran shelters, worked at them, or volunteered at them after Katrina. They were in a no-win situation and, for the most part, did a great job under terrible circumstances.

That being said, post-Katrina shelters suffered from being under-staffed and lacking necessary security, food, water, bedding, medical supplies, hygiene supplies, and toilet facilities. Add to that an incredible lack of support from local government.

Churches offering shelter after Katrina were overrun with refugees that were "unloaded" at their door by city officials. After taking in the refugees, the shelters were on their own. They didn't have radios or any way to contact police or EMS other than sending a runner, and their emergencies had to be prioritized with all of the other emergencies coming in at the time. This meant that shelters had to provide their own security, medical, counseling, food, cooking, water purification, and janitorial services.

As with any population, there were refugees at the shelter who were alcoholics, smokers, illegal drug users, and people who took drugs to control mental illness. I'm not grouping these people together for any reason other than the fact that they all shared the common link of increased irritability when the substances that their bodies were dependant on were no longer available.

The result was fighting, people throwing chairs through stained glass windows at churches, toppling statues, defecating on floors and walls, and a general lack of control.

In one case in Mississippi, SWAT teams had to be called in to restore order because people who ran out of medications required to keep their condition under control began to be a threat to themselves and others.

Another serious problem with the shelters is universal. Any time you put a large group of people in close proximity, add stress, lack of sleep, and inadequate bathroom facilities, you're going to have an increase in sickness. This only compounds the existing problems.

If you are in a disaster situation where shelters have been set up, try to avoid them at all costs, unless you are choosing to help at one.

Once you enter a shelter as a refugee, you will lose your belongings, weapons, and any tools that the shelter workers think you could use to defend yourself (or hurt them). The sick irony is that you may not get any more food or water than if you stayed away from the shelter and fended for yourself. You'll also have to put up with HORRIBLE attitudes, whining, people with entitlement mentalities, and your sleep will likely be interrupted by snoring, arguing, kids blasting music, crying, the threat of violence, and other interruptions that you wouldn't have if you simply holed up on the street somewhere with others you trust and took turns sleeping.

After Katrina, shelters were so overrun that many didn't have any record of who was there or who had been there. Families who were split up for intake sometimes got bussed to completely different shelters, with no record of where their child/spouse/elderly parent was sent. The lesson here is that whatever you do; don't put yourself in a position where you voluntarily get separated from your family or group.

The areas surrounding the shelters were also very dangerous. This is where a certain contingent gathered, including criminals who refused to give up their weapons, people who got kicked out, and drug dealers looking for customers. In a survey of 100 drug users who were in shelters after Katrina, most said that they had no trouble finding their drug of choice within one block of the shelter.

Ironically, most pot smokers said that they stopped using after the storm so that their thinking wouldn't be impaired.

If you live close to a structure that has been declared a shelter area after a disaster, you should seriously consider relocating to a safer location as soon as possible...preferably to the home of someone else on your mutual-aid team.

I'm From The Government. I'm Here To Help.

Centralized approaches to disaster relief failed miserably after Katrina. As an example, there were over 100 helicopters flying over New Orleans within 48 hours of landfall. They were flying 24/7 saving people, but FEMA still hadn't given agencies the green light to start flying and was actually telling first responders to stay away. The 100 helicopters were the result of Coast Guard, National Guard, and other helicopter owners completely ignoring FEMA and deciding to just take care of business.

More insidious was Mayor Nagan's order to disarm citizens and force residents to leave their homes. These next couple of paragraphs are very difficult for me to write, as most of my close friends are either military or LEO.

For the most part, the Army, and various National Guards respected people's rights to stay in their homes. News reports and videos on YouTube reported that the National Guard DID force people from their houses, but the actual videos that are labeled as "National Guard" actually show local police entering the houses.

At least one National Guard officer appears to have told his troops to use intimidation, including drawn weapons, to "encourage" people to leave their homes. However, they weren't authorized to actually force people out. I don't agree with this stance, and it troubles me that the officers in charge appear to have blatantly ignored the Constitution, but for the most part National Guard units did a good job of restoring order.

With the prevalence of "hack job" reporting and video editing, this may or may not have occurred. Again, videos of Guard soldiers and police were often spliced together in Michael Moore fashion to create stories in the viewers' minds that may or may not have actually happened.

There are videos online that take comments out of context that make Guard troops look very bad. I encourage you to dig a little bit every time you see a video like this. After watching about 6 hours of YouTube videos on Katrina, I was able to quickly recognize sound bites that were taken out of longer interviews. In one case, a soldier was asked about shooting fellow US citizens. In the sound bite version, his response was set to music and they make it appear as if he was looking forward to shooting Americans. In the full interview, he was lamenting about how he'd rather be in Iraq where the threat was insurgents and how much the thought of being forced to shoot another American troubled him.

I AM defending the Guard troops. It is an emotional, visceral stance that I take and I want to be clear about it. If anyone sends me substantiated, verifiable information about Guard troops, I will update this chapter, the resource page, and include it for future students. Two thousand students have taken this course at this printing and none of them have sent any videos or anecdotes…substantiated or otherwise.

Unfortunately, law enforcement was another story. There is much dispute about how widespread law enforcement abuse was, but there is no doubt that it happened. There is one widespread story of a frail lady named Patricia Konie who was assaulted, forcibly removed from her dry, stocked home, sent to South Carolina and not allowed back to her home for a month. While the officers who assaulted her were California Highway Patrol officers, local law enforcement was complicit in covering up the offense and sending her out of state.

Local law enforcement also entered homes and disarmed residents, stopped cars and disarmed occupants, and stopped boats and disarmed passengers. In a few cases the firearms were returned undamaged, but in most cases, they were returned damaged or destroyed, or not returned at all. In at least one case, the firearms were beaten against the pavement until the guns were inoperable. I've got a link to a video interview with some of the people who had their guns taken on the resource page.

I encourage you to watch it yourself so that when you talk with people who doubt that anything really happened like this after Katrina, you can tell them with conviction that gun confiscation DID happen, that the 2nd Amendment was violated, the 4th Amendment (search & seizure) was violated, and the 5th Amendment was violated (due process). People weren't arrested and taken to a prison where they were guarded and fed...they were simply detained and disarmed. Watching the interviews will also give you real life situations that you can use to game your response in case you are in the same situation.

If you want to keep your guns in a post-disaster situation, here are three of the actions you can take if you are confronted by anti-self defense/anti-constitutional law enforcement that don't involve conflict:

1. Live in a state that prohibits gun confiscation when a state of emergency has been declared or push for the legislation to be passed in your area.
2. Lie, hide your guns, and/or give one (a decoy) to the police and hide others, etc. This may or may not be illegal.
3. Get on the other side of the law by joining your local Sherriff's auxiliary, search and rescue, joining your local CERT (Civilian Emergency Response Team), or getting licensed for Armed Private Security in your state. None of these are guaranteed to give you a pass when you're dealing with an Orwellian law enforcement official, but they will improve your odds considerably.

Another strategy in this vein is to team up with the family of a law enforcement officer so that you can claim that you have your weapons to protect the family of a fellow officer. Again, there is no guarantee that it will work, but you will want to exploit as many emotional triggers as possible to find the one(s) that will allow you to stay armed.

Thank God For Bubbas And Red Necks

Shortly after word got out about the levee failing, New Orleans was quickly swarmed by people from the bayous and swamp country in shallow bottomed boats. It wasn't a coordinated effort, they just all knew that people would be in need in the city and they wanted to help. Some of these Cajuns barely spoke English and most people from the coasts would call them "Bubbas," "Red Necks," or "Hillbillies" out of contempt. I'd call them that too, but I'd call them that out of respect and shared values.

These guys were heroes. They rescued people who were stuck in trees, stranded on rooftops, and trapped in attics. They didn't need anyone to tell them what to do…they just went out and took care of business.

After any disaster, you'll likely see the "Bubbas" and "Red Necks" coming out of the woodwork to help those around them. They're good people to befriend, both before and after a disaster. Ironically, these good Samaritans had a hard time finding shelter, food, water, and fuel. Most of the hotels that were operable in New Orleans had a policy of not letting ANYONE in after the storm. It made sense, because they would have been overwhelmed if they'd started letting people in, but it is still unfortunate because they turned away volunteer rescue workers as well.

If there is a local disaster in your area and Bubbas and Red Necks start showing up to help, do what you can to help them, whether it's giving them a safe place to sleep on your back porch, food, water, gas, or anything else you can spare without sacrificing Operational Security.

Criminals Will Come Out of the Woodwork

Books, TV shows, YouTube videos, and independent films have been made about the horrible crimes that happened after Katrina. I'm not going to detail them here. There were car-jackings, boat-jackings, murders, racial attacks (in all directions), robberies, smash and grabs, rapes, beatings, etc. Unfortunately, none of those were uncommon for New Orleans before Katrina, and they're still quite common today. Basically, thugs did what thugs do when they think they can get away with it…they steal things and hurt people.

Withdrawal Sucks

One way to capitalize on this is to stock up on cigarettes, chewing tobacco, and/or Nicorette to barter with after a disaster. Keep in mind that the average smoker smokes 15 cigarettes a day and that withdrawal symptoms are worst during the first week and are exacerbated by stressful situations. After Katrina, 52% of smokers increased their smoking by more than half a pack a day on average.

If you're a smoker and you think this is cruel, go stock up on cigarettes or stop smoking now so you don't have to go through withdrawal in a disaster situation. I don't mean it to be cruel...it is what it is...a legal way to stock up on something that will likely go up in value after a disaster that has an illogical, emotional component to it. As a bonus, cigarettes can't be used as a weapon on you (like bullets,) make the person violent (like alcohol), or take away from your family (like food). Cigarettes will also always have barter value where dollars, silver, or gold may not be practical.

Keep in mind that as high as 15% of the general population is on anti-psychotic drugs. That means that several people taking this course will either be on anti-psychotic medication or live with people who do. Some people will be taking the drugs for social reasons, others for minor problems, and a small percentage take anti-psychotics so that they don't have to be hospitalized. To those of you taking anti-psychotics, I can't emphasize enough how important it is to have as much medication on hand as possible. You know how you are off of your medication and a disaster is not a good time to be at anything less than 100%.

If you are able to eliminate any drugs that you're currently taking without serious negative effect, you will be better off. I want to strongly encourage you to consider going to the Mercola Clinic in Chicago to see if they can help you control your condition without prescription medication. The URL for the clinic is: www.naturalhealthcenter.mercola.com. If you can't get off of your medications, then have a talk with your doctor and try to get "ahead" on your prescriptions, even if it means paying out of pocket.

Riding Out The Disaster

This is, in part, the theme of the entire course, but there are a few quick lessons that were made particularly clear after Katrina.

- Scavenging is not looting. Some people may disagree with me on this, but I consider "looting" to be an act of personal enrichment or taking someone else's survival provisions. I have created a two-question test that I will use in urban survival situations: 1. Is there any chance that the owner will return and need this item for survival? If no: 2. Do I need this item for survival? So, the only case when I would take something that wasn't mine would be if the owner had obviously abandoned the item and I needed it for survival. Otherwise, I would not take someone else's belongings.

  I spoke with two private security contractors whose daily routine included finding abandoned cars and siphoning the gas from them to power the generator at the hotel they were protecting. It was relatively easy to do without harming anyone…they simply found a car that had been destroyed by a tree, knock on the door of the house that it was in front of (if it still had a door) to make sure that there weren't people still living there, and take the gas.

  This is much different than stealing food from a house where people are living, taking wide screen TVs from appliance stores, or stealing gas from drivable cars that people are using.

- Regardless of the disaster, if you retreat to your attic, have a way to cut through your roof if exiting through the house may not be possible. Several dozen people were trapped in their attics after Katrina for 2-3 days. Some died, and one man survived 18 days trapped in his attic. I have a link on the resource page for a roof hatch called the "Katrina Hatch" that may be a good solution for you if you live in a hurricane-prone area (or it may inspire you to create something similar).

- Generators attract looters. UrbanSurvivalStories.com tells about one thief who stole 36 generators after Katrina. He would take a lawnmower on low idle in the middle of the night, put it by a home generator, idle it up over the course of 10 minutes or so, and then take the generator without the homeowner knowing what happened. There are a few solutions to this problem, including chaining your generator, using dogs as a sentry, running the generator in your garage and venting the exhaust to the outside, installing or fabricating a muffler, buying/fabricating a generator cabinet that is secured and helps muffle the noise, or improvising an alarm system.

- Adults with special needs, children, and the elderly all got transported to general shelters without identification, medical information, contact information for relatives, or required medications. This happened to hundreds of people and was not isolated to a particular city, race, age, or organization.

  In some cases, the medications were seriously needed. When the patients' drug levels started dropping and they began showing signs, the personnel at the shelters had no idea what they were dealing with. Violent encounters occurred, including at least one where SWAT needed to be called in to restore order, and the situation could have easily been prevented with Medic Alert bracelets or dog tags.

  If you, an adult relative, or someone you take care of becomes dangerous to themselves or others or non-communicative when they are off of their medications, consider getting them to start wearing either a medic alert bracelet or necklace, if they aren't already. In addition to their name and condition, it should have your name and emergency contact information. Try to have both a primary and secondary means of contact on the tag(s), if possible.

Protect Your First Responders

One of the biggest problems that we've seen after recent hurricanes, earthquakes, floods, the 9/11 attacks, and the Oklahoma City bombing is that first responders may accomplish superhero feats, but they are still human and subject to sleep, food, and water requirements. In addition, they have families that they worry about during any downtime that they get.

If you're not currently a first responder, there are a few things that you can do to help the people who are:

1. Get medical training. CPR/First Aid at a minimum and EMT/OEC/WFR training if possible. Any treatments that you can provide to yourself, your family, your team, and your neighbors will take the load off of the EMS system and centralized treatment areas.
2. Get CERT training. This will give you a broad base of post-disaster skills that, again, you can use to help yourself, your family, your team, and your neighborhood.
3. If you have or intend to get firearms, take formal training to learn how to use them. The more you can do to be able to take care of yourself and your family, the better. Said another way, the more self-reliant people there are in a city, the lighter the load will be on first responders.
4. As you get to know first responders who live close to you, ask them if they would like you to look after their family/pets/house in the event of an emergency. Most families who have addressed survival issues have only one key person. In first responder families, this one key person will likely be taking care of others rather than being able to take care of their own family. Your help will allow them to mentally "check off that box" until they get the rest of their family up to speed and could be a welcome relief.

   There is an assumption that neighbors will take care of each other, but having something definite in place will allow their sub conscious mind to relax. While they are risking their life to help others, won't need to be thinking, "Why am I here…my family needs me."

5. If first responders are operating in your area after a disaster and you don't have the skills to help them, offer them food/water and ask them if there is anything else you can do for them. You could even go door to door in your neighborhood asking for people to pitch in. This would build a sense of community as well as possibly give you good intelligence on how your neighbors are doing and what supplies they have.

Get to know the disaster protocol for first responders in your area. One smaller jurisdiction near a large city that I spoke with has the following plan in place.

When a state of emergency is declared, ALL first responders are to report to work and are expected to hot-bunk until the state of emergency is over. They are expected to have their families trained and able to take care of themselves. A local school will be available for all first responder families to go to, should they wish. Shelter, cooking supplies, water, fire, security, and basic medical care will all be provided for first responder families so that the first responders won't have to worry about them.

Police, fire, & EMS are expected to have the same attitude as if they were in the Guard/Reserve, got activated, and got deployed for an unknown period of time.

More common is the following, especially in larger jurisdictions: *Anyone who is working will continue working for the duration of the emergency. Anyone who is not working will first get their family squared away and then report for duty for the duration of the emergency. Families are on their own.*

If you live in an area that is small enough to go with the first plan, you may want to see if local authorities would be open to you helping them put such a plan in place in return for being able to take part in it during a disaster situation.

If you are serious about implementing a plan like this in your area, please contact me. The jurisdictions that have taken this approach are pretty tight-lipped about it and I may or may not be able to put you in contact with them. The decision is not up to me, but I will help you to the extent that I am able.

If you are a first responder and already have a protocol in place, please let me know what it is (anonymous is fine) on the resource page or through email.

To Do: Start identifying first responders in your neighborhood and talking with them as time/opportunity permits. When you feel that it is appropriate, mention how you were reading about police, fire and EMS after Katrina, how many of them were worried about their families while they were working. Ask if they have a plan in place for taking care of their family while they are working during/after a disaster. If they don't have one, tell them that you would be more than happy to help them out if it is ever necessary.

Chapter 11 Resource page:

www.urbansurvivalplan.com/483/katrina

Chapter 12. Psychology for Survival Situations

What is the biggest single factor between success and failure in a survival situation?

It is not going to be the number of guns you have, how much ammo, food, or water you've amassed. It's not even going to be the courses you've taken, skills you've mastered, or the logistics that you've nailed down. It's going to be your mind.

The mind is truly the cornerstone of survival. We talked a little bit about it early on in the course and now we're going to cover it more in depth. I waited until later in the course to cover this important topic for a few reasons, the main one being that the skills are harder to measure and progress is harder to evaluate.

I will be giving you some concrete exercises that you can do to make your mind more resilient in this chapter, but they are much harder to measure than checking off a box when you've got your 72 hour kit ready. I'm going to be mentioning and recommending some additional books that will help you prepare your mind, not only for TEOTWAWKI (the end of the world as we know it) scenarios, but also for many of the problems that are coming on the horizon.

With our spiraling national debt, China threatening to stop buying our debt, $600 TRILLION in derivatives credit default swaps spiraling towards implosion, and the impending commercial real estate crash, not to mention our money supply doubling since 2008, GDP dropping and unemployment rising, we are looking at UGLY economic times ahead.

A confession…

Although you probably signed up for this course to prepare for a post-Katrina or post-terrorist type events, you may not have known that it would help you prepare for the economic freight train of hurt and pain that is currently coming toward us at full speed.

I don't have a crystal ball and I would be speculating to tell you exactly what will happen, but I am not speculating when I tell you that we have some very rough times ahead of us. I have written every chapter with the understanding that while natural and manmade disasters are POSSIBLE, the economic disaster that we're facing is PROBABLE.

With that in mind, this chapter may be the most important one of all.

Burning Ants With A Magnifying Glass

Do you remember playing with a magnifying glass? Do you remember burning ants, blades of grass, paper, and match heads with the focused beam? Do you also remember looking through the magnifying glass and seeing objects go from very small to very large as you moved it?

I want you to think about that for a second. When you look at an object through a magnifying glass, the object does not change size, but your perception of the object DOES change. In some cases, the object is clear, at other times it's fuzzy, and still at other times it is completely inverted.

Take a mental leap with me for a second and think about how you deal with problems and situations. In a very real way, problems and situations are like objects that you're viewing through a magnifying glass. In this case, the magnifying glass is your mind. You have the power, with your mind, to make problems larger than they really are, smaller than they really are, clear and understandable, blurry and confusing, or completely irrational (inverted).

The reality does not change. I'm not proposing that you can stop bleeding with your mind or change who's in control of the country with a wiggle of your nose.

But you can control your response to them. And the more control you have over your mind, the better your response will be.

This applies to shoot/don't shoot situations, determining whether or not someone at your door is a friend or foe, situational awareness, your ability to lead/follow, your tendency to avoid injury, your ability to avoid delusional thoughts, and basically every other thought process you have when you're under stress.

> **To Do**: Think about something that made you very upset recently. Analyze the incident and your response to it. Did you magnify the severity of the situation? Did you give it too little importance? Did you stay level headed? Was this a conscious process or was it automatic?

How important is psychology in stressful situations?

As Lt. Col. Dave Grossman points out in his incredible book, "On Combat" in WWI, WWII and Korea, there were more psychiatric casualties than there were soldiers killed, and the numbers were about equal in Vietnam. Psychiatric casualties were cases where soldiers were unable to continue fighting. This was due to "shell shock," post traumatic stress syndrome, "battle weariness," or any of a dozen other fancy titles for burnout.

This wasn't because they were weak, pansies, or of "poor stock." In many cases, it was exactly the opposite. They were gung-ho and willing to give 110% without recharging until they had nothing left to give. By that point, they were like a car battery that had been drained too low…they needed a complete overhaul in order to function properly again.

In a survival situation, you may not have the option to leave the "front lines" for a few days of R&R to recharge your mind & body.

In other words, mental resiliency is very important. Developing a strong mind will help you during calm times, but is essential to your survival if/when we have an economic collapse, after disasters, and during extended breakdowns in civil order.

So, how can you make yourself more mentally resilient?

The first answer is so simple on the surface that it is often overlooked in search of "sexier" answers. It's sleep. As Patton said, "Fatigue makes cowards of us all."

When you think about it, it makes sense. Everyone accepts that if you don't consume enough calories for an extended period of time, your performance will drop. When told, most people accept the fact that you can survive without food for longer than you can survive without sleep. Even so, few put as much importance on getting enough sleep on a daily basis as they put on getting enough food. Don't believe me? Look at the number of tired, overweight people you see every day.

How Long Can You Survive Without The Following?	
Calm Mind	Heart attack in as fast as 4 minutes
Air	4 minutes
Water	4 days
Sleep	Worthless after 4 days
Food	4 weeks

I've heard sleep called the "great equalizer," but I haven't always believed in sleep. I've tried every strategy I could find to sleep less, including buying books and using supplements, lights, or habits to "add 1-3 hours of productive time to my days." What I found is that if you want to maintain a high level of performance, you can't overcome your body's need for sleep for any appreciable length of time.

Modern militaries of the world have invested millions of dollars over the last several decades into trying to figure out ways to let troops stay awake for longer periods without a drop in performance and they haven't been able to do it. Drugs work for a short time, but in order for soldiers to stay sane and coherent, they just plain need to sleep.

As Col. Gregory Belenky, lead sleep researcher at the Walter Reed Army Institute of Research said, "Warfighters will encounter a lot of information and need to be able to process it at their level to make decisions. You can have a brilliant plan, but unless you have intelligent execution at the lowest level, it won't work." "Sleep-deprived warfighters, because their higher order thinking is impaired, can cause accidents and 'not so clever decisions.'"

There are heroic stories of soldiers fighting for several days with little to no sleep, but they all experience a drop in performance and will burn out for an extended period or permanently if they aren't given a chance to recuperate.

As Army Field Manual 22-51, The Leaders' Manual For Combat Stress, states: "The sleep-deprived soldier or leader has difficulty thinking and reasoning and becomes easily confused and overly suggestible with poor judgment. Pessimistic thinking takes hold and everything seems too difficult. Sleep loss alone can cause the tired brain to see things which are not there (visual hallucinations) or to perceive things which are there as something totally different. When anxiety and vigilance (staying awake on watch) are added, the soldier may be temporarily unable to distinguish between reality and what he fears. Normal physical symptoms of stress can become magnified into disabling illnesses."

The US Army did a study that is particularly applicable to our study of urban survival. They had four mortar crews conduct drills with varying amounts of sleep for 20 days straight. The study is applicable because it showed the impact of sleep deprivation on soldiers who were performing both mental and physical tasks.

Some of the lessons from these tests are in Appendix G of US Army FM 7-90. I have a link to it on the resource page so you can read it in it's entirety, if you are interested. Here is a chart of the hours of sleep that they received and their effectiveness at the end of 20 days:

Group	Hours of sleep per 24 hours	Effectiveness after 20 days
I	7	98%
II	6	50%
III	5	28%
IV	4	15%

> **To Do:** After looking at the chart above, make a conscious decision about what level of effectiveness you want to operate at and make a note of the number of hours of sleep needed to achieve that effectiveness.

When you think about the fact that these are highly-trained soldiers doing an activity that they are experts in, it really shows how important sleep is for urban survival. You probably do not train for urban survival extensively every day. The stresses will be new. The skills that you'll be using every day are skills that you've only used occasionally in the past. As a result, your need for sleep will be as much or more than these highly-trained soldiers.

Near the end of the exercise, group IV was given the coordinates of a hospital that was marked on their map. They completely skipped the step of checking the coordinates on their map and "shelled" the hospital. (It was on a training range and there wasn't really a hospital there.) They had been given the same coordinates earlier in the exercise and did not fire, but in their exhausted state, they couldn't remember a simple step in a sequence that they'd done hundreds of times.

In addition to a drop in mental and physical performance, a lack of sleep causes you to be more irritable, heal slower, gain weight, learn slower, and it causes you to age faster.

> **Note:** I'm very excited about this chapter and its potential to impact your life IMMEDIATELY. If you're currently getting six or fewer hours of quality sleep every night, you can double your mental & physical effectiveness over the next 20 days for FREE by simply getting 7 hours of sleep per night.
>
> If you think that you don't have time, think again. The one hour difference between sleeping 6 hours and 7 hours per night DOUBLED productivity in the remaining 17 hours of the day.

FM 7-90 also talks about the effects of going days without any sleep: "A rule of thumb is to expect a 25 percent degradation (drop) in performance for every 24 hours without sleep. Under the extreme demands of combat, units historically have conducted sustained operations for a maximum of 120 hours. The result was a total deterioration of combat." The drop-off is faster for someone who is hungry, thirsty, out of shape, or who has multiple adrenaline dumps (high stress incidents).

Two other Army field manuals have particularly good information on sleep deprivation. In particular, FM 22-51 talks about the fallacy of people who think that they can operate on little to no sleep. "It is commonly thought that adequate levels of performance can be maintained with only 4 hours of sleep per 24 hours. In fact, after obtaining 4 hours of sleep per night for 5 to 6 consecutive nights a Soldier will be as impaired as if he had stayed awake continuously for 24 hours."

How impaired is someone after staying awake for 24 hours? Fortunately, I've got that information for you. A research project was undertaken from 2001-2007 called "Fatigue and its Effect on Performance in Military Environments," which you can reach from the resource page. In summary, 24 hours without sleep is equivalent to a .09% blood alcohol content (BAC) and 48 hours without sleep is equivalent to a .15% BAC. As of 2005, all 50 states have BAC limits of .08% for driving. To view this chart, go to the resource page for this chapter: www.urbansurvivalplan.com/502/lesson12

As an aside, I must say that there are people, including my sister-in-law, who have operated on 4-5 hours of sleep per night for decades. They are definitely outliers and this is a genetic anomaly rather than a goal to try to achieve.

In "On Combat," Lt. Grossman explains what happens to most people who think they can get by on 4-5 hours of sleep per night when they are put in the timeless environment of a sleep lab. With rare exception, they start out by sleeping 12 or more hours per 24 hour period and once their sleep deficit is made up, they level out at 7-9 hours of sleep per 24 hour period.

Naps

We're going to take a look at the "band-aid" (naps) before we look at the permanent solution (good sleep.) If history is any guide, the reality of survival is that it is an exercise in compensating and making due. You're not always going to be able to get the ideal amount of sleep, and so you need to know how to compensate with napping. Naps do make a significant difference in how long you can go without sleeping.

Please go to the resource page to view a chart from Col. Gregory Belenkys' study "Sleep, Sleep Deprivation, and Human Performance in Continuous Operations." : www.urbansurvivalplan.com/502/lesson12 It shows the mental performance over time of people who took one 30 minute nap per day with no other sleep. As well as the mental performance over time of people who did not sleep or nap. After as little as 3 nights without sleep, the person who took three 30 minute naps is operating at almost twice the level of the non-napper.

Insider Powerful Napping Secrets

Here are some quick do's and don'ts for getting the most out of your naps:

1. Find a quiet place where you won't be interrupted. Wear earplugs or listen to white noise if necessary. Stay away from radios, TV, phones, and conversations. Give yourself the freedom to not respond to any stimuli.

2. Schedule your nap. Make sure that people around you know you are taking a nap and that you aren't to be bothered. Accept the fact that you're going to be napping, identify the benefits, and let yourself do it.

3. Go somewhere dark, wear eyeshades, or otherwise cover your eyes. I like to lie on my back and put my arm on my face so that my bicep covers one eye, and my forearm covers the other eye. I use my other hand to support my neck.

4. Get in a comfortable, stable position so you don't have to flex muscles to stay there.

5. If you're lying on your back, either put something under your lower back or lift your knees and put your feet on the ground so that you don't wake up with a sore back.

6. Avoid caffeine before napping.

7. Naps will be most productive when your body is at its natural low points throughout the day. These occur in the morning, just after lunch, and at night.

8. Set an alarm or ask to be woken up somewhere between 5 and 20 minutes.

9. Take a few deep breaths as you're settling down for your nap.

Three personal tricks that I have found to help my naps have more impact:

1. I have found that simulating rapid eye movement while napping seems to help me wake up more refreshed. This is purely anecdotal and I have not read of any studies being done, but it works well for me. I simply shut my eyes and dart them back and forth, up and down, in geometric shapes, and randomly. If someone looks at me while I'm doing it, it looks like I'm in REM (rapid eye movement) sleep.

2. Many times, I get a significant benefit during napping from simply letting my mind digest all that I have been throwing at it so far that day. Other times, as thoughts (troubling, to-do's, etc.) enter my mind, I imagine that they come in the left side of my head, I put them in a box, and put them on a conveyor belt that goes out the right side of my head.

3. Sometimes, when life is particularly hectic, I find that praying and turning everything over to God is the only way that I'm able to calm my mind enough to nap.

If you're very tired, you will probably experience "sleep/nap inertia" when you wake up from your nap. Simply put, it means that you wake up groggy. This effect takes longer to overcome when you take naps that are longer than 20 minutes. Two quick ways to help get through this phase is to have some caffeine upon waking from your nap or exercise briefly.

Strategies for Effective Sleeping

Keep in mind that everyone is different and you may require more or less sleep. The scientific way to figure this out is to go to a sleep lab and sleep until you don't need sleep anymore. That's not practical for most people, so start off by recording the following:

Date	Time to sleep	Wake Time	How Long?	# times awake	Refreshed next day?
1/1	2300	0600	7 hrs	2	Yes! Very.

Try to go to sleep and wake up at near the same time every day for a couple of weeks and see if you feel better.

If, after a week and a half, you find that you can't sleep until 0600 anymore and that you're consistently waking up at 0530, it probably means that you need about 6 ½ hours of sleep per night. More likely, you'll find that you still need an alarm clock to wake up at 0600 and that your sleep requirements are more than 7 hours.

If this is the case, go to bed an hour earlier for a week or two and see how you feel.

I have found through trial and error that I can get by on 7-8 hours of quality sleep per night, but that my body performs best on 8 ½-9 hours. I've also found that I need an additional hour of sleep (for a total of 10 hours) during allergy seasons to keep going at full speed. Now that you know my sleep requirements, you can probably better understand why I've spent so much time and money trying to figure out a way to sleep less.

> **To Do:** Take the next two weeks and schedule 8 hours per night to sleep. Keep a sleep log with columns like the ones above.
>
> If you don't think you have time for 8 hours of sleep, realize that it's likely that you can DOUBLE the effectiveness of your waking hours by getting the amount of sleep your body requires.
>
> Testing among Olympic athletes has shown that they must have 7 ½ to 8 hours of sleep per night for optimal results.

How to get more sleep with fewer hours in bed

Just because you allow for 7 hours of sleep doesn't mean that you'll get 7 hours of high-quality sleep. There's a huge difference between high-quality sleep and low-quality sleep. Most people can think back to a night when they drank too much, had coffee too close to bed time, took cold medicine and woke up repeatedly, slept in an uncomfortable bed, or spent the night in a hotel room next to a loud party.

These are all examples of times when just spending time in bed doesn't translate into feeling refreshed the next morning. If you want to get the most benefit from your time in bed, it's important to focus on the details of sleep. Here's how:

1. Get your sleep in one uninterrupted period, if possible. The peak regenerative period of sleep happens between hour 6 and hour 7 for most people. You could call it the grand finale of sleep…the rest of the show may be good, but the tail end is amazing. There are obvious problems with this if you are in a survival situation and you need to set up multiple watches. Until then, try to follow this guideline.

2. Get as much of your sleep between 2200 and 0600 as possible. Again looking to a survival situation and multiple shifts, one solution would be to have one shift sleep from 1900-0200 and the second shift sleep from 0200-0900.

3. Get your sleep at the same time every night.

4. Control the conditions where you're sleeping. Pick a place that is dark, quiet, temperate, and safe. Wear earplugs, eye shades, and adjust clothing if necessary.

5. Exercise will help you sleep better in two major ways. First, it will help your body "burn up" stress hormones that would otherwise keep you from getting deep sleep. Second, it will allow your mind to deal with stresses easier. In a post-disaster situation, 20-30 minutes of calisthenics (Especially turbulence training and/or kettle bells) every day will help calm your mind and allow you to start napping and sleeping quicker, and stay asleep longer.

6. Self-permission to sleep. Don't sleep with one eye open. Trust the people you are with, trust your dog, or trust your intrusion devices, but whatever you do, create a situation where you can give yourself permission to sleep.

7. No alcohol, big meals, or excessive sweets 1-2 hours before bed. All of these will cause insulin/adrenaline cycles that will impair sleep. Alcohol may make you drowsy, but when it's converted to sugar, it will hurt the quality of your sleep.

8. No caffeine for 5 hours before bed. It will inhibit the release of melatonin, keep most people from getting to sleep as quickly as they would otherwise, and will make your sleep less recuperative than it should be.

9. Make up sleep deficits as soon as possible. US Army guidelines state that soldiers should go no longer than 2 weeks on 4 hours of sleep per night.

10. Calm your mind. Many people drink or take pills to "calm the demons" so that they can sleep. The problem with these approaches is that they hurt the quality of sleep. Instead, I suggest prayer, tapping (link to info on the resource page), exercise, and the strategy I mentioned about "putting stuff in a box" and letting it pass through your head. Another good way to do this is to practice "making a baby" ☺

11. Stay hydrated. If your urine is yellow in the evening before going to bed, you need to drink more water. Without getting too involved, when you sleep, your blood vessels dilate and your blood takes oxygen and nutrients to your muscles and carries off waste. When you are dehydrated, this process is impaired and your muscles won't be as effective the next day. You may have to wake up in the middle of the night to urinate, but most people who experience this problem will find that their sleep is so improved by being hydrated that it is worth it.

12. Eat smart. Try not to go to bed hungry, but also avoid eating more than a snack within two hours of going to sleep. You particularly want to avoid large quantities of foods high in saturated fat, sugar, and simple carbohydrates.

13. Avoid TV, surfing the internet, video games, or other forms of escapism when you're trying to sleep. They don't regenerate the mind or body, and in many cases keep the mind from dealing with the situations that are at hand. There is a big fear among the Killogy Group (Grossman, De Becker, Watt, & more) that our soldiers in Afghanistan and Iraq are going to have worse PTSD than in past wars because they are spending their off-time "escaping" rather than sleeping. As a result, their mind never has a chance to properly process what it's seen and done.

Quoting Lt. Col Grossman from "On Combat" again, "Many law enforcement agencies have told me that they have a serious stress problem and want me to teach them how to deal with it. When I tell them they must first ensure that everyone gets enough sleep, they argue that it is impossible. 'Okay,' I say, 'Then die.' I am kidding, of course, but I say it to make the point that if they want to draw overtime, enjoy a full career and watch their grandbabies grow, then one of the most important things they can do is manage their sleep."

In other words, take this seriously. Sleep provides a solid foundation for a solid mind, dealing with stress, and making good decisions.

Exercise

I've been fortunate enough to become friends with one of the advisors for Killogy, Lt. Col. Randy Watt. Randy is one of the top firearms, edged weapons, leadership, and situational awareness instructors in the country and I'm going to be putting together some training packages with him that I know you will benefit from. Randy wrote a creed that I have adopted as my own called, "One Warrior's Creed," and I have included a link to it in the resource section for this chapter.

Randy was one of the key influences for this course and one of the things that he stresses in all of his classes and that he stressed in every interview with me was the importance of exercise. It affects EVERY aspect of your life. Sleep, energy level, your ability to deal with stress, firearms accuracy, martial skills, and clear thinking to name a few.

I have to add a disclaimer...make sure to check with your health care professional before starting any exercise program. You could injure yourself or die.

That being said, no matter what your age (I know of a few 80-year-olds who have completed this course) or your physical ailment, find out what physical exercises you can do and start doing them. If you're confined to a wheelchair and can't use your legs, then see if you can start doing arm exercises to get your heart rate up for 20-30 minutes a day. If your hands are inoperable, or if you have spinal issues, ask your doctor or a therapist what you CAN do to exercise and get your heart rate up, but whatever you do, figure out a way to get exercise.

For those of you without physical ailments, I've included links to two of my favorite exercises on the resource page that will work even if you're a desk jockey. They are Pistol One-Legged Squats and One Armed Pushups. I like these exercises because they help you build up your supporting muscles and you can easily vary the intensity. Both of the links show Craig Ballantyne from TurbulanceTraining.com doing the exercises.

The pistol one-legged squats can be modified by squatting your butt all the way to the floor (hardest), squatting your butt down to a chair (as shown in the video) or by squatting your butt down to the arm of a couch, a window sill, or a desk (easiest). Whichever intensity you choose, I suggest only doing 5 at a time with perfect form multiple times a day. As a fine point, as you do them, make sure that your lower leg stays perpendicular to the ground and that you can see your toes over your knee the entire time.

The one-armed pushups can be modified by doing them with your knees on the floor to make them easier, by pushing against a chair or desk, rather than against the floor to make them easier or by raising your feet to make them more difficult. Again, I suggest doing multiple sets of 5 with perfect form throughout the day. Perfect form is KEY with one-armed pushups, just as it is with the pistol squats.

I find that by doing these exercises (and pull-ups, using a hang-board in my office) throughout the day, I stay fresher and more awake than if I do nothing or use caffeine.

I'm also a BIG fan of kettle bell workouts, trail running, and mountain biking. If you have any questions on these exercises, feel free to contact me by email or through the forum.

Other strategies for keeping your mind under control

- Pray and read your Bible. There are few things more comforting than knowing that God is in control. If you don't believe in God, that's fine. All of us who do will enjoy our delusional stupor. ☺

- Accept the fact that fear is OK. Only delusional people will operate without fear in a survival situation. It's much healthier to identify fear, accept it, and use it as a high-performance fuel. The more you can learn to identify and LOVE the feel of fear-induced adrenaline coursing through your veins and use it to improve your performance, the more you will become the master of it.

The best way to do this is to put yourself in controlled situations where you get adrenaline rushes. Not everyone can stack up and raid a house, serve high-risk warrants, or set up and execute an ambush, but there are other activities that you can do to help you get used to adrenaline. Some of the "safe" activities I enjoy that most people can easily do are sparring, shooting competitions, rock climbing, rappelling, high-consequence negotiating, and public speaking. Other things like skydiving, crawling through a tight tunnel, getting a hood thrown over your head, handcuffed, and thrown in a trunk, or something as simple as climbing on a roof may give you a rush.

One of the stress inoculation drills that I have done in Krav Maga is one where 6 people gang up on me while I'm lying on my back spread eagle. One person grabs each foot, presses pressure points, lightly applies ankle locks, and rakes their forearm across my shins. Two more people grab each arm, and alternate between kamoras and key locks, and hitting pressure points. A fifth person alternates between doing pushups and digging their elbows into my stomach. And the sixth guy works my head, rubbing my eyes, plugging my nose, covering my mouth, choking me, and hitting pressure points.

It is a GREAT drill to go through. The point isn't to escape, although I try. The point is to stay calm, not panic, and to get used to the adrenaline dump.

The important thing is to identify the rush, accept it as your body's high-quality survival fuel, and learn how to deal with it. Famed firearms trainer Massad Ayoob talks about a time he was competing at the Bianchi Cup and all of the competitors had a case of the butterflies. While other people were bouncing around, he calmly did deep breathing and isometric exercises. In particular, he brought his foot up and held it, stretching his quads. As he did this, he pushed outwards with his foot and inwards with his hand. He did this with both legs and effectively stopped his leg shakes before shooting his stage. He didn't try to suppress or get rid of the fear/anxiety/stress/adrenaline. Rather, he calmed his mind and heart and harnessed the adrenaline to shoot a great stage.

To Do: Plan an activity that will give you an adrenaline rush. (You must consult with your healthcare professional to make sure that you can handle it).

Things to remember

- Give yourself time to mourn. In a survival situation, you WILL lose things. It could be your comfortable life, possessions, friends, or a family member. You don't want to make your life about focusing on what you lost, but your mind will need to work through various emotions as you are moving forward with your life.

- Keep looking forward, keep moving forward, and don't dwell on the past. Many people were paralyzed after Katrina with anguish over what they lost or fear of what they were going to lose. From the interviews that I did, it appears that these people were useless to themselves and useless to their friends and families. The people who thrived after Katrina and who are doing the best mentally years later are the ones who made a conscious effort to refuse to be a victim.

- Set SMART goals (Specific Measurable, Achievable, Realistic, Timely) for every individual and for the group. It could be as simple as making sure everyone gets 7 hours of sleep, that you get to spend 15 minutes reading your Bible, and that you talk with 3 of your neighbors every day for the next week. In a survival situation, you might have a map of the area and set a goal to explore 10 new blocks every day until you have covered every street within a mile of your house. If you have 100 gallons of water, 6 people and 2 dogs and you're figuring 2 gallons per person per day and ½ gallon per dog, you will want to have a goal to secure additional water for your group within a week if you don't have a way to recycle it.

One of the big benefits of SMART goals is that they give your mind something to focus on. Your mind WILL focus on something. It can focus on poor-little-'ol-me, or it can focus on something productive. Give it productive things to focus on and it will serve you well.

To Do: Practice setting SMART goals. As an example, you could say, "When I pay off my Visa, I will take a survival or firearms course." You may have to modify the reward to something like, "…we will take a trip to a city where I can take an Urban Escape and Evasion class and my wife can spend 3 days at the spa/pool."

In a survival situation, it could be, "Today, I need to find a source of water, transfer 20 gallons to our location, and make it potable. When I am done, I will read a fiction book for 20 minutes."

Write down five SMART goals right now. They can be for your survival plans, work, relationships, relocating, or anything else. Adding a reward will help you achieve the goal, but is not necessary.

- Only allow yourself to think about things that you have control over. Worrying about what the weather is going to do will not help you. Creating a hasty plan for what you will do if various weather events happen WILL help you. Worrying about which of your rights the government might eviscerate next will not help you. Getting active politically by volunteering, helping a candidate or elected official with research, or by donating can help. As worries or concerns enter your mind, evaluate whether or not you can do anything about them. If not, think about something else. If it is something that you can do something about, come up with a hasty plan, write it down if necessary, and expand on it later if it makes sense to do so.

As a personal example, when my brother and my friends are deployed, I find myself worrying about them frequently. My worrying doesn't do me or them any good, and it actually harms me. Instead of worrying, I've developed the discipline of praying for them, writing them letters, or sending packages. These actions all make me feel better and they help my brother & friends.

-Smile, joke, laugh, and play games when possible. Laughter refreshes the soul, and in a survival situation, you're likely to need lots of refreshing. Besides light hearted laughter, know that you'll probably find yourself laughing at inappropriate things. Medical professionals, firemen, and soldiers often laugh at disgusting, sad, and gut wrenching situations.

There are many stressful situations where the line between crying and laughing is blurred. Someone watching from the outside might think that you are being irreverent, but in reality it is a simple coping mechanism that our brain uses to give us an emotional outlet. Just accept it, go with it, and know that most people will not understand the humor, but hopefully the people that you're with will.

-Have confidence. In God, in yourself, your team, your equipment, your training, the future, and your ability to improvise, adapt, and overcome.

-If you have a choice between being positive and being negative, be positive. It doesn't cost any more, and it will improve your performance and the performance of everyone around you. That doesn't mean that you shouldn't anticipate problems. You should. When you do, figure out a solution, move on, and improve on it as time allows.

Never waste time focusing on problems when you can use that time to come up with solutions.

It is subtle, but the difference in results is dramatic. Focusing on problems is degenerative in nature. It eats away at your mind, your sleep, and your relationships. When you identify a problem, immediately focusing on solutions trains your mind to look forward, to anticipate a better future, and to spend time on thoughts that can actually help you.

To Do: The next time you "consume" news or talk radio, consciously monitor your emotions, breathing rate, and heart rate. It doesn't have to be exact…just know whether you are breathing faster or slower than normal.

Analyze your reaction to each story and then decide whether or not there is anything you can do about it. If there is nothing you can do about it, tell yourself, "That's unfortunate. I wish I could do something about that, but I know that worrying about it won't help anyone. Fortunately, I can do a set of 5/10/30/50 pushups right now so that I'm in better shape and better to handle stress in the future."

If there IS something you can do, do it immediately or schedule it immediately so that you can let it pass through your mind.

"That's unfortunate. I am going to write my representative an email RIGHT NOW to tell them what I think." When you're done writing, you can say "I've done what I can for now."

Or

"That's not good. That's too big for me to fix, but I can pray for God to fix it."

The point of the exercise is to practice controlling your reaction to the news so you can apply that control to other areas of your life. As a good friend pointed out to me, most things that cause the most extreme worry and "mental paralysis" are things that are out of our control. By identifying that fact and letting them go, you'll be more able to change the things that you do have control over.

These skills will be easier for some and harder for others. Some people are predisposed to worry and be anxious while others could care less about what's going on around them. The important thing is to identify where you are, where you want to be, and start making consistent steps to get there.

I've found out the hard way that this discipline is different in men than it is in women. Sometimes women just need to talk about what has happened to them. I don't understand why, even though I've read more than a dozen books on marriage and communication. I just know that most of the arguments that my wife and I have are the result of me trying to fix problems that she just wants to tell me about.

Usually, she's already figured out a solution, doesn't want to hear mine, and just wants to share her experience and emotions with me.

My wife is amazing and understands my need to "fix". One of our secrets is that when she is telling me about a problem and I'm chomping at the bit to help her "fix" it, I will ask her, "Is this something that you want me to help you fix, or do you just want to talk about it?"

Supplements that can help with your mood and sleep.

Not to beat a dead horse, but one of the strongest and most effective drugs that you can take to improve your mental state is the endorphins that are released during exercise.

There are also supplements that you can take that can help your mood and allow you to sleep better.

I am not a doctor and you should consult with a health care professional before taking any of these supplements. There could be horrible interactions with medications that you are currently taking. I'd appreciate the input of any medical professionals on this topic.

A few years ago, I did quite a bit of experimentation with ways to consistently achieve runners' high, how to predictably cause it to happen on command, and how to make it last as long as possible so that I could effortlessly run for hours.

There were three supplements that I used that I found to be helpful in my runners' high testing that I also found to make me more mentally resilient when I wasn't running. They were SAMe, St. John's Wort, and 5-HTP. (Do not take them together unless you are under the supervision of a healthcare professional.)

In addition to helping with mood regulation, SAMe is also very beneficial for joint pain. St. John's Wort has been used as an anti-depressant in Europe for decades. 5-HTP boosts serotonin levels, the natural release of melatonin, and increases the efficiency of your sleep by increasing the amount of time that you spend in REM sleep by up to 25%.

You REALLY need to consult with a healthcare professional before using these supplements. They are seriously powerful and have the ability to raise your serotonin to toxic (deadly) levels when taken alone, together, or with other drugs.

Melatonin is another supplement that my family takes occasionally. You can find it in evening teas, sprays, juices, and tablets, and chews.

You can buy it in .3, 1, 3, and 6 mg doses, but I've found that by taking as little as .25mg and reading, I become unable to read in as soon as 10 minutes and fall asleep quickly. Your experience may be very different, but I've found that I wake up alert and well rested when I take .25mg.

Personally I use melatonin when I travel to the East Coast and want to go to sleep early and get a full night of sleep for early morning meetings.

I also sometimes use it when I have made the mistake of drinking coffee after 1700 in the afternoon. Without melatonin, I will toss and turn for up to 2-3 hours before finally falling asleep. With melatonin, I'm able to go to sleep in minutes.

Read, reread, and study further

I can't emphasize how important psychology is to survival and how vital your ability to positively influence the thought patterns of yourself and those around you will be. In a survival situation, you MAY need firearms skills. You MAY need martial arts skills. You MAY need negotiating skills, but you will definitely need the ability to calm your mind and calm the minds of those around you.

I want to encourage you to read this chapter least one more time. Mark up the sections that resonate with you and study them further. I've included links to several online resources and wonderful books that you can read to understand your mind better and the minds of people around you.

Chapter 12 Resource page:

www.urbansurvivalplan.com/502/lesson12

Chapter 13. Urban Movement After A Disaster

We're going to focus on three major types of post-disaster movement in this chapter:

1. "Just" a disaster with no reason to be covert. People are in shock, no increased risk of violence yet. (Immediately after a terrorist attack like 9/11, unexpected natural disaster, etc.).
2. A disaster with civil breakdown (exploration/gathering/all non-trade) where you may have to switch to covert (economic collapse, predicted natural disaster with pre-planned looting by gangs like Katrina, a breakdown in the electrical grid, food chain breakdown, etc).
3. A disaster with civil breakdown (travel with the intent of shopping/trading) where you need to carry goods/currency and may have to switch to covert.

Thanks to the prevalence of video cameras, we have some great video documentation of peoples' reactions after the WTC towers collapsed on 9/11. Two of the most common problems that the videos showed were people wearing suit coats/blazers while their faces were covered in dust and people running barefoot down the street holding their high heels/ fancy dress shoes in their hands.

We don't have to wait for another terrorist attack to experience the same problem. Hawaii and California experience small tremors on an almost daily basis. Many other parts of the country have seismic activity every week. The majority are too small to feel, but are a constant reminder that major earthquakes can happen. When they do, people will be dealing with debris on the ground and dust/smoke in the air, just like New Yorkers dealt with after the WTC attacks.

The average person spends just under ½ of their time away from home during the week, and it's likely that in a disaster situation, you will start out away from home and you will want to get to your home as quickly as you can. Ideally, you would have your 72 hour kit with you 24/7 and you would always be able to wear running/hiking shoes, but that's just not reality for most people. As a result, we need to be prepared to improvise and adapt if disaster strikes, we're not ideally dressed, and we need to travel through urban areas to get home.

For the point of this section, we're going to focus on a journey from where you're walking to either a "safe house" or your house. The safe house can be the house of any friend where you could go to get shelter, water, food, fire, first aid, possibly communicate, and gear up as necessary to make it the rest of the way home. In other words, we're covering journeys of less than 12 hours where food and water will be smart, but not necessary for surviving the trip.

To begin with, you want to make sure that you can regulate your temperature. In the summer, this may mean turning clothes inside out if the liner is a lighter color than the outside.

In the winter, this may mean stuffing your shirt/pants with crumpled newspaper and/or wearing a trash bag as a coat/kilt. Homeless people use this little trash bag trick, and our family used it growing up when we went to football games on cold rainy/snowy winter nights. (I told you I could identify with rednecks ☺)

If you are in a high dust environment, you will want to filter the air. Dust and smoke in your lungs will keep your lungs from being as efficient as possible. The resulting decrease in oxygen will keep you from traveling as fast and thinking as clearly as you otherwise could. Depending on what chemicals are in the smoke & dust, it could also cause permanent damage to your lungs.

At a minimum, take off a piece of clothing or rip a strip of cloth that you can use to cover your mouth and nose. Try to pick a cloth that isn't too hard to breathe through. A loose knit fabric is perfect, because it will allow you to add or remove layers as necessary.

If you have access to water, wet down your cloth. Water will allow it to catch more debris and breathing the moist air will be easier on your throat and lungs.

Next, we need to protect your feet. When I look at the footwear people wear on a daily basis, I'm amazed at how flimsy most of them are. People regularly go out with ½" thick flip-flops, fancy pointed-toe dress shoes, heels, thin-strappy sandals, and the like. I'm not going to be critical about foot fashions, but if you do choose to wear this kind of footwear you need to know how to take care of your feet if you have to travel long distances in the city.

To share the importance of good shoes for urban travel, I want to share a quick story with you. Like Neil Strauss, I've gone through the OnPointTactical.com Urban Escape and Evasion course and for the practical exam I spent 8 hours evading 9 bounty hunters in a downtown area. I had on GREAT hiking boots. They are actually the third pair of the same model for me. I've wore them for guiding backpacking and mountain climbing trips in Colorado, and I've worn them for rough trail running. I can wear them for a full day of backpacking with a 60 pound pack and still have relatively "fresh feet" at the end of the day.

After 8 hours of walking and evading on concrete in my favorite boots, my feet were shot. We figured that with our pacing, we were averaging about 2-3 miles an hour with stops, and we figure we went somewhere between 15-20 miles.

> Some people have scoffed at the emphasis that I put on quality footwear, saying something to the effect of, "I wore flip flops for 3 days at Disney and I felt fine." This is possible because Disney and many other theme parks use a special asphalt that is impregnated with rubber to keep guests from getting tired and going home early. Concrete (what sidewalks are made of) is 10 times denser than asphalt, and roughly 20-40 times denser than theme park walkways.

The reality is that, unless you have a job where you wear hiking boots or running shoes on a daily basis, you are probably going to need to compensate for your footwear during a long hike in an urban area...even if there is no debris on the ground.

You're going to need to make some judgment calls on short and long term comfort and damage, but here are some general guidelines:

- If you're wearing heels, break them off.
- If you are wearing no socks, thin dress socks, or hose, put fabric or tape between your foot and your shoes to give yourself a little blister protection.
- If you pass a store that carries shoes, buy a pair that will allow you to move faster.
- If you pass a store that carries tape, buy some in case you need to repair or reinforce your footwear. Ideally, buy duct tape.
- If you pass a store that carries superglue, buy some for its MANY survival uses we have discussed already, including shoe repair.

- Walk on asphalt rather than concrete. Walk on dirt rather than asphalt. Walk on grass rather than dirt.
- If you can walk in the street, rather than on the sidewalk, walk in the center of the street. Most streets are "crowned" rather than flat so that water will run off of the street/road. Walking in the middle of the road will be easier on your back. If you can't walk in the middle of the street, try to alternate walking on the right and left side of the street.
- If the bottom of your shoes are very thin, you can reinforce them by putting a coat sleeve, strip of cloth, tape, and any other available material between your foot and the shoe.
- Remember, your best solution is to start with or acquire good footwear. All of these steps are simply ways to compensate for times when you don't have good footwear with you.
- If you have the opportunity, and it is advantageous, use the best mode of transportation you possibly can. i.e. buy a bike, skateboard, rollerblades, razor, etc.
- In cold weather, build up your shoes as necessary to keep your feet warm. Make sure that you don't wrap anything too tight so that you lose circulation to your feet.
- Stay hydrated and snack when possible if you have a long walk.
- As you develop "hot spots" apply duct tape directly to the spot. If a sore or blister is already forming, try to put some paper, cloth, or reversed duct tape over it to protect the loose skin from the tape.

How do I get there from here?

One of my "Franken-soldier" (mercenary/civilian private security contractor) friends introduced me to the concept of PACE routes. PACE stands for Primary, Alternate, Contingency, and Emergency. In an ideal world, you will have PACE routes memorized from your place of work to your house. If you haven't completed that level of preparation yet, you should have local maps in your 72 hour kits.

If you are out to lunch with a group of friends/co-workers when you have to switch into survival mode, you're probably not going to have your 72 hour kit or any of your survival gear with you. This simply means that you will need to be able to improvise and adapt and acquire a map as soon as possible.

In addition to buying a map, two free sources of maps that you're likely to have in almost any city are mass transit maps and maps in phone books. Almost all white pages have maps in them as well as many yellow pages.

> As a personal note, not all private security contractors/private military contractors are evil. Nor are they all good-guys. All of my friends who are PSCs are stand up guys who are still serving their country, even if they are not paid directly by the military. They are red-blooded Americans who would fight martial law rather than enforce it.
>
> They love America, the Constitution, and individual rights. Some of them are still in the National Guard or the Reserves and pick up contractor work between drills and deployments.
>
> Not all PSCs are like my friends, but enough are that I encourage you to do your research before you buy into the demonization of PSCs by the media, Hollywood, and politicians. As a start, I've included a link to a comparison of contractor pay and soldier pay on the resource page. I think you'll be surprised by it. I was.

Urban Travel After Civil Breakdown

The sheer number of potential civil breakdown scenarios makes an exact set of guidelines impractical. If the civil unrest is localized to a few streets, a neighborhood, or a section of town the response will be different than if there is civil unrest across an entire town, which will be different than civil unrest that is regional or nationwide. In addition, your actions are going to depend on WHO is in authority in your immediate area. Is it the police that you know? Are they guard/military troops (who vote conservatively 80% of the time)? Are they UN troops?

In the US, some recent short term examples of localized breakdowns in civil order are small scale riots in LA when the Lakers win the NBA championships and WTO riots in Seattle. These didn't affect people who were a few miles away at all, and people who weren't near the riots when they started could easily avoid them or go around them.

Of course, many situations can't be avoided so easily. Two examples are the power outages on the East coast (short term) and New Orleans after Katrina (short to medium duration).

We could have varying degrees of civil breakdown as a result of similar events, terrorist attacks, high unemployment rates, currency devaluation, food shortages, or other economic triggers.

The guidelines that I'm going to give you are general in nature and you'll want to adapt them depending on your particular situation.

The most obvious trigger to let you know that you are in a situation where you need to follow these guidelines are when the police/fire/EMS are overwhelmed and there is an increase in criminal and civilian violence. Along with those factors, you will likely have one or more of the following:

1. Drinking water shortages
2. Food shortages
3. Breakdowns in power, water, and/or communications
4. Currency instability

Movement away from your house

The movement that we're going to deal with in particular is movement out from your house in a time of extended civil unrest. We're going to assume that the disaster has precipitated a situation where there is a large number of people but a shortage of supplies.

I'm breaking the movement up into two categories: Travel with no intent of trading and travel with the intent of trading. In both cases, we're going to cover travel by foot since it is the most basic form of travel. The biggest difference between the two is the presence or absence of carrying valuables and the increased chance of identifying yourself as a good target.

Why would I want to leave my house?

The problem with survival provisions is that, sooner or later, they run out. The goal with your survival provisions is to have enough on hand to be able to survive until you find a reliable source of consumable items, like food, water, fuel, and medication.

Your survival supplies will allow you to survive without interacting with anyone else for a time, but eventually you'll want to start venturing out and finding ways to start replacing consumables at least as fast as you are using them.

Hopefully, the situation that prompted you to go into survival mode is just a natural disaster or riot and will pass quickly. If not, here are some of the other reasons why you will want to venture out before order is restored.

1. Water
2. Helping friends and neighbors
3. Identifying shelters, gangs, and other threats in your area
4. Finding other mutual aid groups in your area
5. Replacing supplies
6. Finding where medical help is if you need it
7. Gathering intelligence
8. Finding out who is in authority in your area and creating it if there is a vacuum
9. Teaming up and organizing your neighborhood/city
10. Setting, checking, retrieving caches
11. Finding backup shelters
12. Finding PACE routes to your backup shelters
13. Finding PACE routes out of the city
14. In general, finding ways to improve your situation, through better shelter, fire, water, food, or security.
15. Trade/barter/market

As your particular survival situation develops, you're also going to want to know what's going on around you and whether or not it is smart to stay in your current location. As we discussed earlier, if a church, community center, or park that is a block from your house is designated as a refugee center, you may want to relocate as soon as possible.

You're also going to want to know if there are any gangs or community groups operating in your area. Since Katrina, CERT and Neighborhood Watch groups nationwide are preparing for future unrest and will have a presence as soon as an event happens.

Most of these groups will be friendly, but some will be outright unfriendly to outsiders, and others will be living in a narcissistic fantasy land trying to create their own private fiefdom straight out of Hollywood.

As you read survival websites, you'll soon see that this isn't common, but is a very real concern. Some people are HOPING for a disaster so that they will have a chance to be "important" and live out their fantasies developed by watching movies and reading novels. These people are more interested in having instant social status and think that they'll be making all the right decisions like their favorite fictional character. In the event of an urban survival situation, my plan is to identify and avoid groups and individuals with this mindset as they pop up.

Concentric circles of security

You're going to have several zones of security that increase as you go out from your house. As an example, one set of outgoing concentric circles of security could be your safe room, your house, your property, your street, your neighborhood, and your side of a physical barrier, like a river. Each level is a little less secure than the previous one and you need to be more cautious and more aware as you go to less and less secure areas.

At some point, the security level is going to change from one where you're fine walking around alone to one where you will always want to have a buddy with you. Again, this could be your house, or if your street has teamed up, your entire block could be safe enough for you to go out alone. Remember, we're not talking about the day after a flood in Iowa where everyone is helping each other. We're specifically talking about a situation where civil order has broken down.

If you are in a survival situation, you also do not want to leave all of your belongings unguarded. Since you don't want to go out alone and you don't want to leave your supplies unguarded, you can quickly see why it is so beneficial to have at least a few adults on your team, if your house can support it. If not, you should try to team up with one or more neighbors when you go out so that you can go out as a pair and your spouses can look out for each other at home.

Be as boring as you can be

As you're going out progressively further and further from your house, be very mindful of the people around you. Take note of their dress, level of cleanliness, how much they're carrying, whether they're carrying visible weapons, whether they look tired or energetic, whether they are walking with a purpose or listlessly, etc. This is called the "baseline" and you want to match it as closely as possible. As an example, if everyone is wearing civilian clothes, you don't want to go out in BDUs wearing a helmet, a ruck sack, and carrying an M4 unless you want to attract attention.

Another issue with dressing in a "tactical" look after a disaster is that you look more like an authority figure. As a perceived authority figure, you've got increased risk of "bad" people attacking you and "good" people swarming to you for help. If you actually are acting in the capacity of a first responder, than this isn't a problem, but if you're moving about with a purpose, it could become a big pain, real quickly.

To avoid this, try to wear clothes that are as close to what you see other people wearing as possible. Avoid wearing clothing with school, city, or neighborhood names/logos. Conceal any weapons that you're carrying, unless carrying weapons is the norm. If you carry a pack, try to make it one that looks more like a book-bag or courier bag than a tactical bag. If people are moving confidently and with a purpose, do likewise. If they look hungry, tired, and worn out, try to match it.

One of my biggest "tells" in rough neighborhoods is that I'm white, fit, have short hair, and I can have an intense look unless I'm consciously trying not to. In a recent trip to an apartment complex in a seedy part of town, I was walking through the courtyard and a little 3-4 year old girl started pointing at me and yelling, "COP! COP!" while crying and clinging to her mother's leg. It was the middle of the day and I was dressed in a non-descript North Face shirt, shorts, and tennis shoes. She just interpreted my purposeful air (and probably the fact that I was the only white guy within a mile) as law enforcement. If I'd been trying to be invisible and match the baseline, it could have been a big problem.

The baseline will likely be very different as you go from your neighborhood to other neighborhoods. This can be especially true in a survival situation. If people in your neighborhood are well fed, have water, feel safe, and are getting sleep at night, they are going to carry themselves very differently than people in an area that is suffering. As you get a feeling for the different baselines in your area you can plan your appearance so that it is easy to transition from one to another.

To Do: 1. Start identifying the baseline look for the different areas in your city. Figure out what you would do to your appearance to walk through it and blend in. The baseline will likely change in a survival situation, but this exercise will train your mind to start identifying the baseline.

2. Identify the boundary lines where the baseline changes near you. It may be a city boundary, railroad tracks, a highway, a city street, or something else.

What to take?

This is going to be somewhat dependant on what will fit in with the baseline in the areas where you will be traveling. It's likely that a courier bag or a small backpack will be all you can carry externally without standing out.

You can also carry extra gear in cargo pants and, depending on the season, a jacket or coat. Cinch things down and gear up so that you can jump up and down without making noise and having stuff flop all over. In short, try to avoid carrying gear that will be loud or will prevent you from running.

At a minimum, you're going to want to have the following on you: water, snacks, knife, multi-tool, lock pick, lighter, light, comms. After that, it's going to depend on your situation. I would suggest the biggest weapon you can carry without deviating from the baseline, spare mags, a way to purify more water, food, chain and/or a cable, a padlock, a map that does not have any markings on it, a TINY survival/medical kit, pepper spray, small bolt cutters, chalk or a livestock marker, and a notepad & pen.

Why a chain and padlock?

Let me give you a hypothetical scenario. You have a light industrial area near you. From previous exploration, you have identified one facility that is a couple hundred yards long that is abandoned and has 8 foot barbed wire fence, but gates at both ends that are not locked. As you are checking out the neighborhood for threats and resources, you notice 3-4 rough looking teenagers scoping you out.

You happen to be near this facility and duck in one end and speed up to get across to the other end. As you get to the other end, the kids are half way through and gaining on you. As you go through the 2nd gate, you run your cable through the fence, lock it with your padlock and take off.

Now the teens have to either get over the barbed wire fence or double back in order to get to you. You have the option of returning later and retrieving your lock and cable. By identifying this choke point in advance, you were able to easily diffuse the situation and avoid a no-win conflict.

You could also use this technique to block a door by locking the push bar or handle to a pipe, metal beam, or stud. This works better with a cable than a chain, and works even BETTER with big or multiple zip ties, but it depends on the door being solid enough to slow down your pursuers.

The basis for this strategy comes from a fictional story by Victor Aguilar called, "200". I apologize in advance for some of the language he uses, but he has some very good nuggets of information and I'll link to it on the resource page for this chapter.

As a slight modification to this strategy, you could do the following if you had two teams out at the same time. Let's use the same scenario, except this time the gate at one end is locked. Your team and the other team are walking down parallel streets, one block apart so that you will pass by the open gate and the other team will pass by the closed gate. As you spot your tail, you radio to the second team and have them double time it to the locked gate to either pick the lock or cut the lock.

As you enter the open gate, your pursuers will think that they have you trapped and will have no reason to speed up. As long as your other team has defeated the lock by the time you get there, you can walk right through and re-lock it or replace it with your own.

On defeating locks...practice makes perfect. MasterLocks are amazingly simple to rake (a form of picking). They are even easier once you figure out a particular lock. As a novice picking a MasterLock, it might take anywhere from 5 seconds to 5 minutes. Once you know the right combination of wiggles for a particular lock, you will be able to pick the same lock in 1-3 seconds in the future. As you get better, your times will improve considerably.

Keep in mind that picking locks is illegal and that raking damages the pins and will make the keys for the lock not work eventually. In other words, don't pick locks that you want to work in the future with a key. (i.e. the lock on your house, or a lock that someone else is using in a non-survival situation.) But if you have potential chokepoints like this that are padlocked, it might not hurt to "learn the locks" in your area if you are in a survival situation. Just keep your local laws in mind.

This brings up the importance of knowledge about the area around your house and the importance of route selection. The more you can learn about your area now, the better prepared you will be in the event that you have to switch into survival mode.

> **To Do:** look for choke points in your area that could trap you in a survival situation or that you could use to stop pursuers. Mark them on your SurviveInPlace™ map.

What the heck is a livestock marker?

I grew up on a farm and we raised pigs the years when we thought it would be profitable. Whenever we'd work on the pen, feed them, or spend time with them, we'd have a grease marker with us. If we spotted one that was sick, injured, or needed special attention, we'd mark it with the pen so we could identify it quickly when we came back later with whatever supplies we needed. The markers look like BIG crayons and they have grease in them that won't wash away without scrubbing.

Carrying grease pens and/or chalk with you will allow you to use pre-defined "hobo codes" or your own codes to mark points of interest for you or other people in your group when they go out.

What about Murphy?

Before you go out, take some time and brainstorm possible problems that you might run into and pre-plan your response. As an example, what if you don't make it back on time? Can your team/neighborhood spare anyone to look for you? Should they look for you? You could have gotten lost, had an innocent accident, gotten arrested, or you could have gone into a bad area. If it was a bad area, do you want another pair of team members potentially meeting the same fate?

These are questions that you need to ask and answer BEFORE you go out. Some others:

1. What is the appropriate response to local/federal/international law enforcement? The response could be different.

2. How often will you check in by radio? What codes will you use? Will you leave your radios on or will you just have scheduled times to communicate?
3. What is the plan to call you back if something happens at home while you're out?
4. What will you do if one of the pair has an injury that hurts mobility?
5. If you are aggressively pursued, will you split up or stay together? If you split up or are split up, when/where will you meet up? How long will you wait?
6. How will you avoid and detect surveillance?

Detecting and avoiding surveillance

I have sold silver online before to people who live locally. This involved meeting them to exchange silver for cash. Being the "moderately aware" person that I am, one of my first concerns was that they could easily follow me home under the faulty assumption that I had more precious metals there. As a result, I pick public meeting places (I like Starbucks and food courts) and park my car a couple of blocks away. After the exchange, I stick around awhile to read and pick a U shaped route to get to my car so that I can easily tell if I am being followed.

Basically, if my car is Southwest of my meeting location, I will go East, spend a few minutes in a retail store, go South, West, and back North to my car. The chance of anyone following the same route with the same pacing is very slim, unless they are following me.

This is a very basic form of surveillance detection. Two other forms are using reflective surfaces, watching someone who stays at the same distance from you, even after stopping or changing speeds.

Switching to Covert

If you suspect that you are being followed or see someone who you'd rather not recognize you, it may be advantageous to change your appearance quickly. A few simple way to do this are:

- Buy some cheap hair extensions from a drug store or dollar store and superglue them to the inside of a ball cap. Ideally, you want to have this available in advance.

- Reverse or change your jacket. Switch pants and/or shirt.
- Put in colored contacts.
- Put on or change your sunglasses.
- Use fake facial hair.
- Shave or dye your facial hair. (You may be able to quickly shave and look different. I've usually got dirty blonde/white hair. When I shave my goatee/beard, I end up bleeding for half an hour and I don't look that much different than I did originally. I get a much better effect by dying than shaving. You'll only learn this by trial and error.)
- Have alternate clothes/appearance modifiers cached in your area of operations that you can quickly access and change into.

Whatever you do to change your appearance, you need to make sure that you still end up matching the baseline, and not looking like someone at a Halloween party in a cheap costume.

My contacts who have done extensive undercover work, including Kelly Alwood, from OnPoint Tactical, all agree that the only way to get good at it is to practice. Practice will tell you what looks natural, it will tell you which disguises last for a few minutes and which will last for several hours, and practice will give you confidence in your disguise, which will help keep you from telegraphing doubt.

If you get into disguises, one "test" that you can do is to pan-handle at a busy intersection near where you live. For such a safe exercise, the risk of embarrassment makes the "pucker factor" pretty darn high when you see someone you know.

Traveling for Shopping and Trade

Traveling for shopping and trade is split out because at some point during your travel, people are likely to see that you have money or they will see that you bought something valuable. Both of these actions make you a good target.

Before we get into that, let's back up a minute and talk about what items should you try to have on hand for trade and barter.

Ideally, you want to have items on hand that you won't need for survival that you can use for trade & barter. I wouldn't trade clothes, water, food, or fuel unless you have an abundance of the item or a way to replace the item easily.

Some examples of items to trade are multi-tools, cigarettes, novels, small games, soap, anti-diarrhea medication, sugar, salt, 550 cord, or, as suggested by James Rawles from SurvivalBlog, high-capacity magazines that you own specifically for trade. Ideally, you want consumables that are cheap now and that people would pay a lot for if there was a shortage. Another GREAT item is matches or lighters. Ragnar Benson tells about urban survival situations in other countries where people went through 750 matches PER MONTH!

Of course, if you've "picked the right horse" and have the accepted currency on hand, you can simply use it. What I mean is that the accepted currency in a medium to long term urban survival situation could be FRNs (Federal Reserve Notes/dollar bills/paper money). But, it could also be "junk" silver (pre-1965 US coins), silver 1 oz rounds, modern US coins, or even gold for large purchases. The fact is, the medium of exchange can be whatever a particular society decides that it is.

It could also be salt, bullets, cigarettes, gas, or something completely different. After Katrina, dollars, fuel, water, and food worked as currency. When Argentina had their massive inflation, coins were worth more than paper money. In Zimbabwe, they knocked 12 0's off of the paper currency and kept the coinage the same...essentially ADDING 12 0's to the value of the coins. They eventually started using foreign currency to conduct business. We just don't know what the currency will be until the disaster is upon us.

My suggestion is to buy some of each with an emphasis on small denominations of cash, junk silver, multi-tools, cigarettes, and other consumables. If you LOVE knives, water filters, solar/crank gadgets, or something else that you never seem to be able to buy enough of, then buy new versions with the intent to sell/barter/trade the old ones eventually.

Money Management (How to keep from having all of your money stolen)

If you've traveled much overseas, you're going to be nodding your head in agreement as you read this section. The reason is because you're going to use the same anti-theft strategies in a post-disaster survival situation that you use when you're traveling in a foreign country, in a bad part of town, or, if you're like me, in everyday life.

First of all, we want to get into the mind of a thief or a husband/father of a hungry family. For the most part, thefts are crimes of opportunity and you can do a lot to prevent them by not exposing yourself as a target or making yourself look like an easy mark.

I love watching animal shows. Not the stupid pet trick shows, but the wildlife shows where a lion will stalk a herd of gazelles, looking for dinner. The lion knows that she isn't going to eat unless she kills and she has to use her resources efficiently or she and her cubs will die of starvation. As a result, she doesn't go after the largest gazelle...she goes after the one that she has the best chance of catching and killing without getting injured or using up precious resources.

When you're traveling to and from your trading destination, you will be a ripe target. You'll have all of the elements needed for survival AND you'll have money.

Getting back into the mind of the hungry husband/father/thief; he is going to look for signs that you are worth robbing and evaluate the potential of getting hurt/arrested/killed. We've already covered the baseline look, so your dress/appearance isn't going to make you stand out and the next thing is where/how you carry your money.

You always want to carry your money in front of you...in your jacket, shirt pocket, neck wallet, money belt, front pockets, cargo pockets, shoes, etc. so you can use your eyes to help detect thieves. The most common form of pick-pocketing is called the "bump" and "pick" and takes advantage of this fact. It's very simple. One person bumps or distracts you and a second person picks your wallet. This is easiest with a "checkbook" wallet or what one of my private military contractor friends calls a "George Costanza" wallet, but it can be done with any wallet.

Think about the last time that you paid cash for something. Did you reach in your pocket, pick out the right bills by feel, and hand the cashier exactly the right bills? Probably not -- unless you are blind. More than likely, you pulled out your wallet or your cash, sorted through it, and pulled out what you needed.

As you did this, it's likely that you telegraphed how good of a target you are.

What bills were visible to the cashier? To other shoppers? Keep in mind that a fat wallet looks valuable, even if it's full of $1 bills and discount cards. Think about how many "gold" or "platinum" cards you flash when you pay for things. All of these are telltale signs that you may be a good or bad target.

But what about the "blind" crack you made?

Many blind people actually CAN reach into their pockets and pull out exactly the right currency to pay a bill without pulling out all of their money.

In England, they can do it because the bills are different sizes. In Canada, the bills actually have Braille on them. But in the US, they do something different.

Since US bills are all the same size, they do it by using a system of folding. $1 bills are left flat, $5 bills are folded lengthwise, $10 bills are folded by width, $20 bills are folded lengthwise AND THEN by width.

This allows them to tell what bill they're holding without being able to see it. It also allows them to reach into their pocket and pull out no more than what they absolutely have to when they're making purchases.

When I'm traveling or at fairs/markets, I carry cash in different pockets. I also do this in Vegas when I'm playing blackjack.

I'm a card counter and basic money management dictates that you should always have 30X your base bet. If I'm playing at a $25 table, I've got at least $800 on me, plus any winnings for the day and my spending money. One way that I split it up is as follows: I put my spending money in my left front pocket. I put $400 in each of my cargo pockets and my winnings in my right front pocket.

This way, when I want something to eat/drink, I can reach into my left front pocket and pull out "all of my money" and it's only $40-$50. When I sit down at a table, I reach into one of my cargo pockets and pull out everything that's in there and lay it on the table. I don't count out $400 from a bigger wad of cash…I already know how much I'm laying down as soon as I reach in my pocket. If I want to put down $500 instead, I pull out $400 from one pocket and a single $100 from the other cargo pocket. Since I know that the only thing in that pocket is $100 bills, all I have to do is pull out the first bill I grab. With practice, you can do this so that it looks like that's all that you have in that pocket.

I also use a two step process. It is handy when we're at theme parks, in the winter, when I'm skiing, or when I'm doing security/close protection and have my pockets full of "goodies". It's very simple. I start out with a single $20 in my pocket and more $20s in a neck wallet. As I get close to using up the $20, I will either pull out another $20, or if that doesn't seem smart, I'll go to the restroom, enter a stall, and pull another $20 from my neck wallet. This way, I can just pull "all" of my cash out of my pocket without ever showing very much money and making myself a target.

You can use any of these methods (folding, putting a single denomination in a pocket, or two stepping) to effectively hide exactly how much money you've got. The exact method is not important...rather, focus on the concept of dividing your money and hiding how much money you're carrying. If you know that you're going to buy something expensive, know how much it is going to be, and can get that much money isolated before you buy it. I like to pre-position the exact amount of cash in my left front pocket so I can pull it out without having to open multiple pockets.

Splitting up your money like this also allows you to empty 1, 2, or even three pockets in the event of a mugging and still not get wiped out. If I'm in a rough area, I go so far as to put an old hotel key and a couple of non-activated "loyalty" cards in my back pocket with some cash. If I need to, I can grab them and toss them at a mugger and use the distraction to either attack or run.

That's all well and good for cash, but if you are trading or bartering, you will probably need either a shoulder bag or a backpack.

Use the same strategies with whatever currency or exchange medium you are using...pack the items into various "denominations" so that you don't have to ask people to make "change" for you. This will help you get a better deal and it will help keep you from showing off how much you actually have with you.

Keep in mind that the process of negotiating, buying, and exchanging currency for another item will give people in your immediate area an idea of how much you are carrying with you. If there are very unsavory people in your immediate area, you should weigh the benefits of completing the purchase, asking the vendor to do the deal in private, waiting until they leave, or waiting until another time.

Strength in Numbers

Again, the best way to avoid becoming a target is to make sure that you don't look like a target. But what if you have to make several small purchases in the open or you need to make one large purchase? The easiest solution to both of these situations relies on having a team with you.

If you have several small purchases to make, split them among the group. If you have one large purchase to make, you can do it in private. If you have no reason to think that anyone sees you as ripe targets, than just act nonchalantly and head home. Stay aware, but look nonchalant.

If on the other hand, you feel like you were "cased" or are being followed, you can approach the purchase as if it is a "close protection" detail where the buyer or the most likely target is the "principal" and the rest of your team is the protection detail (bodyguards).

Essentially, have everyone but the principal spread out and looking around for people who are paying too much attention. You still want to stick to the baseline look, but be observant of what is going on around you and who is paying attention to the purchase.

There are two ways to approach this after the purchase:

1. You have made yourself an obvious target. In this case, you will want to travel in what is called a "tight diamond" formation. You will have one person 5-15 feet ahead of the principal, 5-15 feet behind the principal, and one person with the principal (assuming 4 people). If you have additional people, they will take the same relative formations, but further out. The goal here is to make yourself look like a big gazelle. When they see that there are 3-4 people obviously protecting the principal, they know that they will probably sustain losses if they try anything.

2. You have not made yourself an obvious target, but you feel like being cautious. As you leave the store/market and start heading back home, you want to use what is called a "loose diamond" formation. Essentially, you'll have one person 10-30 yards ahead of the principal, a second person either with the principal or across the street from the principal and a third person 10-30 yards behind the principal. Additional people can be tighter, pair up, or be further away from the principal. Done correctly, a loose diamond formation is almost undetectable in

a crowd. It's just a few people going the same way on the same day.

In general, the more of a threat that you perceive, the more formidable you want your group to look. The less of a threat that you perceive, the more you want your group to blend in.

In any of these scenarios, you will want to go through the same surveillance detection routines (reflections, turns, pacing) as we discussed earlier.

I mentioned something at the very beginning of this chapter that's worth repeating. Everything in this chapter is easier with a bike, vehicle, horse, or the use of mass transit. I've illustrated everything with the worst case scenario (only foot traffic) since the lessons would, by default, cover the easier scenarios.

You've probably noticed a lot of overlap in these sections...baseline, choke points, observing your surroundings. That's because these are all fundamental skills. The more you practice them in the coming weeks, the better equipped you will be to survive and thrive in both non-survival and survival situations.

Assignments for this week:

To Do: Identify 3 choke points near your house, work, or in between that could trap you or that you could use to evade others if necessary.
To Do: Consciously identify the baseline look around you as you go through your week. Create their "story" in your mind. Do the same with people who obviously stick out.
To Do: See if you can identify the boundaries where the baseline look changes.
To Do: Try one of the money tricks mentioned to minimize the amount of cash that you expose when you make purchases.

<u>Chapter 13 Resource page:</u>

<u>www.urbansurvivalplan.com/516/lesson13t</u>

Conclusion

Congratulations on completing the Urban Survival Guide! It is a topic that I hope you can tell I am very passionate about and is close to my heart. I firmly believe that the future of our country depends on the number of self-reliant people...QUICKLY. Please recommend this book and my other urban survival preparedness items to your friends and relatives.

Remember, it was written to be digested in weekly lessons and installments. If you have not completed the tasks and assignments, please go back and do so. Also, remember even if you have completed them to go back through the material so that you will be able to add to your level of retention. As I stated previously, your retention of the materials covered have been shown to go up 9X by completing the exercises. If survival is important to you, it's worth taking the time to lock the information into your brain.

Following this is an appendix with web pages that are resources for you to use going forward with your Urban Survival Plan. I have made these pages so that they can be updated and I can continue to give you the best information I come across. I have also included my twitter ID so that you can reach me if you have questions.

If you haven't yet, go join our forum at: SecretsOfUrbanSurvival.com so that you can be aware of information as it becomes available and so you can learn from other like-minded people.

Also, make sure to sign up for Urban Survival updates by filling out the form at SurviveInPlace.com/book When you do, I'll immediately send you out a guide for clearing your own house with a firearm.

God Bless,

David Morris

SurviveInPlace.com/UrbanSurvivalGuide.com
UrbanSurvivalPlayingCards.com
SecretsOfUrbanSurvival.com
Facebook.com/SurvivalDave
Twitter.com/SurvivalDave

Resources

Below you will find a list of web pages with resources for each of the chapters. Some are referenced in the Chapters. They are set up as web pages so that they can be updated as new information becomes available.

Urban Survival Guide blog and forum:
www.SecretsOfUrbanSurvival.com

Urban Survival Playing Cards: (52 tips, tricks and secrets to keep you alive after an urban disaster) www.UrbanSurvivalPlayingCards.com

Forum:
www.SecretsOfUrbanSurvival.com

Chapter 1:
www.urbansurvivalplan.com/595/lesson1

Chapter 2:
www.urbansurvivalplan.com/592/lesson2

Chapter 3:
www.urbansurvivalplan.com/590/lesson3

Chapter 4:
www.urbansurvivalplan.com/588/lesson4

Chapter 5:
www.urbansurvivalplan.com/193/lesson5

Chapter 6:
www.urbansurvivalplan.com/306/lesson6resources

Chapter 7:
www.urbansurvivalplan.com/320/lesson7t

Chapter 8:
www.urbansurvivalplan.com/365/househardening

Chapter 9:
www.urbansurvivalplan.com/382/lesson9

Chapter 10:
www.urbansurvivalplan.com/441/lesson10

Chapter 11:
www.urbansurvivalplan.com/483/katrina

Chapter 12:
www.urbansurvivalplan.com/502/lesson12

Chapter 13:
www.urbansurvivalplan.com/516/lesson13t

Chemical attacks:
www.urbansurvivalplan.com/299/saferoom

Situational awareness:
www.urbansurvivalplan.com/49/sipbonus

Surviving Inflation:
www.urbansurvivalplan.com/535/surviving-inflation

Overheated RE & Stocks:
www.urbansurvivalplan.com/544/realestateandstocks

Follow me on twitter: www.twitter.com/SurvivalDave

Made in the USA
Lexington, KY
30 December 2010